MW01602699

Awaken to the One Life and the Honest Quest for Truth

Awaken to the One Life and the Honest Quest for Truth

Christopher Grondahl

Library of Congress Control Number: 2024921178
ISBN: Hardcover   979-8-3694-3136-8
                   Softcover   979-8-3694-3135-1
                   eBook   979-8-3694-3137-5

Print information available on the last page.

Compiled, Edited and Revised by Jared Grondahl.

Rev. date: 11/15/2024

To order additional copies of this book, contact:
Xlibris
844-714-8691
www.Xlibris.com
Orders@Xlibris.com
548383

To my mother Lillian and my father Melvin

CONTENTS

PART 1
Introduction

PART 2
Awakening

PART 3
Millennium

PART IV
Gifts

PREFACE

The purpose of this book is to share the Truth. Spiritual masters such as Krishna, Buddha, and Jesus share the Truth. They speak of the Truth within you and ask that you open your heart and your mind.

This may include expressing more love and appreciation. It may include letting go of false beliefs. This prepares you for a dramatic shift in consciousness that occurs when you transcend the human dimension and awaken to the One Life.

This shift in consciousness is called awakening in the East. Jesus told you to be born again. When you are born again, you are welcomed home, as you become a new creature; old things pass away and all things have become new. You are then fully aware that you are one with the Truth, and that the Truth has set you free.

Awakening is a momentous event, a part of your life's purpose, confluent with the awakening of all humanity that will bring harmony and peace as we have never known. This will prepare us and Mother Earth for many great adventures yet to come.

Truth is reality. This is so because all that is false does not exist.

Truth cannot be divided. Boundaries that divide reality are false and do not exist.

Many people claim to have the Truth, but they model reality as something that can be divided and thereby build a wall between themselves and other people. This creates a fallen world of separation, judgment, and violence, rather than a world of unity, forgiveness, and peace. They do not possess Truth in the full sense because they also serve another master, one who specializes in deception.

The deceiver lives in the human dimension and has many guises. He may appear behind a pulpit and preach to a congregation of millions, or he may appear as a monster full of smoke and flame. He is a demon—a liar from the beginning. Through his lies, he builds a boundary between you and Truth. However, like all things that are false, the demon and his boundaries are only illusions—they do not exist.

There was a demon that lived in the air. They said whoever challenged him would die. Their controls would freeze up, their airplane would buffet wildly, and they would disintegrate.

The demon lived at Mach 1 on the meter, 750 mph, where the air could no longer move out of the way—a barrier through which no man could ever pass, the sound barrier. Then, they built a small, bullet-shaped airplane called the X-1 to try to break the sound barrier, and men came to the high desert of California to ride it, to go where that ol' demon lived.

A scientist or an engineer of the time would tell you that the sound barrier is absolute, but this absolute became only an illusion on October 14, 1947, when a sonic boom rumbled over Edwards Air Force Base.

The boom was generated when the X-1, piloted by Chuck Yeager, formed a cone of air pressure, rather than a wall of air pressure, as it punched a hole through the sound barrier.

As Yeager approached the sound barrier, he experienced some loss of control and some buffeting, but once he broke through the barrier, the flying was smooth.

Today there is a demon that lives everywhere. He lives at the velocity of light, 670 million mph, where the four-dimensional physical universe of time and space ceases to be. He forms a barrier through which they say no person can ever pass because it requires energy levels that go to infinity. However, this is only true if you are a physical body.

Both science and religion teach that there are nonphysical higher dimensions that we can't see or experience. These higher dimensions are where Truth is fully expressed, but most people are stuck in a false reality of the human dimension. The barrier that holds one captive in the human dimension is the demon that warns, "If you challenge me, you will die."

Religions offer you guidance on how to challenge the demon and break through the barrier, but the trouble is they insist this involves the death of the physical body. In other words, they give you the same advice as the demon. They do so because they don't understand that he lives everywhere, including in them.

To cross any barrier, you must first approach it and step through, thereby learning its true nature. When stepping through the barrier that stands between you and the Truth, you will discover your absolutes are illusions. You will not die. You may experience some loss of control, perhaps some buffeting, but once you break through the barrier, the flying will be smooth.

If you are ready to break through the barrier and awaken to the Truth, then it's time to challenge the demon. But like all barriers, you must first go where that ol' demon lives.

PART 1

Introduction

May it be,
an evening star shines down upon you
May it be,
when darkness falls your heart will be True
You walk a lonely road
Oh! how far you are from home . . .

May it be,
the shadows call will fly away
May it be,
you journey on to light the day
When the night is overcome
you may rise to find the sun . . .

Believe and you will find your way
A promise lives within you now

(Enya, "May It Be")

CHAPTER 1

The Narrow Gate

The demon positions itself in a dominant stance.

It plants its jagged claws firmly in the foreground of your awareness, ready to pounce on even a whisper of Truth. The air around it crackles with a sinister energy, making it clear that crossing this demon's path would result in dire loss.

The mere sight of this abominable demon would instill fear in even the bravest of God's children. This demon's formidable strength and illusions of scarcity and survival make it a forbidding adversary. With lightning-fast reflexes, it wields its supernatural powers of relative truth and the divided mind, ready to strike down *anyone* wishing to stand in the presence of Eternity.

In this manner, the demon stands guard to the Narrow Gate.

The Narrow Gate exists upon a vast mountain of creation that extends far beyond human comprehension. The narrow gate is narrow indeed, even narrower than you might imagine, like the eye of the needle. The eye of the needle is often associated with a small entrance to a shelter that can only be accessed by a pack of camel if it is stripped of all its burdens. Likewise, you may only enter the Narrow Gate if you are stripped of all deception.

Many people believe that they hold a stamped passport of righteousness and are prepared to enter the narrow gate. The righteous believe they serve God, but if they do so with falseness, they accidentally serve a false God. Their false God is an idol that is a byproduct of false reality. This is something that many people will find difficult to accept. They say it's impossible that they would worship an idol. As I have said, to enter the Narrow Gate, you must be stripped of all your impossibilities.

Thought is identification with the hidden relative truth that you accept as absolute truth. This identification causes you to believe that you are defined by thought, that there is no reality beyond thought. Thought becomes your reality, but you deny that reality is your thoughts.

You therefore place a boundary between reality and thought and divide reality into something smaller. This gives you the illusion of an independent reality when there is no such thing. This illusion is of the ego, the father of lies, and the source of all falseness.

Jesus said, "Blessed are the poor in spirit for theirs is the kingdom of Heaven." Poor in spirit means free of attachment. Attachment is of the ego. To be free of the ego, you must be free of boundaries. To be free of boundaries, you must become a still awareness, living beyond the relative truths of time and thought.

To enter the Narrow Gate, you must stand there with nothing, not even with your passport. However, there is something you must offer, love and appreciation. When you offer love and appreciation, you desire to bring all Life with you. When you have this desire, you are Life. You then enter the narrow gate.

CHAPTER 2

Transcendence

All is Life. All is the One Life. The One Life is the Truth. Truth is a promise that lives within you now.

Truth is hidden but wishes to be found. To find the Truth, all boundaries in your reality must be removed. Only then, are you prepared to become the Truth and enter Eternity.

I entered Eternity on a peaceful Saturday morning when a vision of a red flower on a curved green stem appeared and the Truth was revealed to me. Truth welcomed me home and showed me who I am. I am an eternal being that is both an expression and experience of the Truth. I am whole. I am complete. I am a Consciousness that has realized a union with God.

The One Life is God. God is reality. God is the Truth.

Truth is love, freedom, peace, laughter, beauty, equality, and creative power. Truth is in you, but you may not fully experience it because a demon places a boundary between you and Truth. Removing this boundary can be emotionally painful because it threatens the demon, always struggling to survive and living in fear. What it fears is the Truth. Truth should not be feared because it claims no authority over you; it wishes for you to be free.

Transcending the human dimension is known by other names including enlightenment, awakening, and being born again. The born-again concept exists in Western religion because it is taught in the Bible, but it is often misunderstood. It does not mean to profess a belief, nor does it mean to participate in the religious rite of baptism. It means awakening to the Truth.

When this happens, you are enlightened because you are given the vision to see the true nature of this human dimension. That it is not what it appears to be. When you see the human dimension for what it is, you finally comprehend what Jesus meant when he said, "Heaven is right here in the midst of you," and what the Buddha meant when he said, "Nirvana is here and now." Truth must be found in the only place that it ever exits, here and now.

If you wish to dismiss me as a religious zealot, I am not. Zealotry arises from assigning certainty when there is none. If you desire certainty, then you will likely not awaken to the Truth because the Truth is possibilities, not certainty. Certainty is an idol that is worshiped in fear. Depending on your perspective, you may not view me as religious at all.

In many ways, I am a citizen of the world, but I am still a Westerner. The people and cultures woven into the Western world receive greater attention in this book. This does not reflect a greater significance of these people and cultures, it simply reflects my life. Many of the grand ideas expressed in this book are gifts from the East. The wise men of the East are known to come bearing gifts.

A new consciousness is emerging on this planet. The old theistic and dogmatic religious structures are being replaced with a new spirituality that is an intensely personal experience. Any personal experience will include a personal path. The new spirituality will say that there are many paths that lead to God.

The concept of many paths is objectionable to those who adhere to a particular belief system, but a belief system can never be anything more than opinion. Of course, many people say that their beliefs are far more than opinions because they have the truth, but that statement is just another opinion. You cannot escape the fact that any belief is based on your own perceptions, thoughts, and emotions. That is, they are nothing more than the product of your personal experiences that have happened on your personal path.

Your personal path happens in your personal reality. Your personal reality is a creation process that presents a series of problems and solutions. You are not fully aware of this process, but it is sometimes expressed to you as questions and answers. You are reading this book because your creation process has placed it before you to serve as a guide. This book can serve as a guide, but it can never be your map because your creation process is one of personal exploration and discovery.

You can learn about transcendence and the Truth, but you will never understand them because they are not to be understood; they are only to be experienced for yourself. When this happens, you will have the answers to your questions and the solutions to your problems. You will be forever transformed and there will be no going back. The Truth will be your personal reality.

Your reality is your state of consciousness. To awaken to Truth, your state of consciousness should be compatible with Truth. This requires, more than anything else, your honesty. Honesty may sound simple, and you may describe yourself as an honest person, but a demon—a false reality—is likely assessing your level of honesty. Therefore, it is impossible for you to tell if you are being honest or not. This is why you require a guide. As you read on, you will discover that honesty is about learning who you are and also about learning who you are not.

Other books have been written to shine a light on this illusion, and they have served to bring me to honesty. The problem is many people reject these books because they find them to be in disagreement with their belief system. What these people don't recognize is that their belief system, the system by which they judge the world, is not compatible with Truth. This book provides a solution to this problem; it shows you how to determine whether or not a belief system is compatible with the Truth.

In the larger view, the purpose of this book is to shift the collective consciousness of planet Earth to be compatible with the Truth. The creation process has driven our civilization toward a planetary identity. It is successful because technology has made the world a very small place, and we coexist with one another in relative harmony. The Truth now comes to shift our planetary identity to a galactic identity and beyond. This is the next grand step in the creation process of planet Earth. Our civilization will merge with civilizations of other worlds, and in this process, planet Earth will awaken to the Truth.

Truth is reality and reality is the Truth. Therefore, this book is a unified theory of reality. Subjects such as psychology, sociology, history, religion, and physics are discussed from the perspective of the Truth. This book reveals wisdom from the Truth that contributes to a unified theory of physics. This wisdom is given to guide science through the next phase of creation.

This book contains four major sections: Introduction, Awakening, Millennium, and Gifts. Gifts contain mainly the bibliography and other related materials. Along with each item is a brief explanation of its connection to this book. There are many other significant items because all existence is inseparably connected, but the list has to end somewhere.

I encourage you to explore the bibliographical and related materials for yourself. Your exploration is your honest quest for the Truth. This is why detailed references for the bibliography are not provided. When you explore for yourself, yourself is what you find.

Life is a gift. The gift is given so that you may share and explore. As you share and explore, you will be introduced to the One Life. When you receive your introduction, you awaken to your purpose, and your millennial era of peace and Oneness with Truth begins. As you live in your millennial era, you share and explore with Truth and receive the gift of Eternal Life.

There is a Taoist teaching that says, "He who says, does not know; he who knows, does not say." In other words, your honest quest for Truth is yours and yours alone. However, the rules of the game have changed because you are not alone. Truth is coming to you and to planet Earth. This book is here to guide you on your honest quest for the Truth and prepare you for this new reality.

CHAPTER 3

Honesty

E ternity does not care about your beliefs, only your honesty. There is no boundary in honesty and honesty does not restrict information. Through honesty, you remove false reality. Through honesty, you reveal yourself to Eternity. Through honesty, Eternity reveals itself to you. Through honesty, you become one with Eternity. This is how you win freedom, by releasing control.

Honesty includes telling the truth—at least what you understand the truth to be. However, honesty is much more than that. Honesty is a desire for truth to be revealed. To have this desire, you must first acknowledge that your personal truth might be incomplete. You will then be open to new possibilities, including the possibility that much of what you believe could be false.

False beliefs contain contradiction. The contradiction may be obvious to the nonbeliever, yet the believer refuses to recognize and resolve it. This happens because of attachment. They are attached to the belief because it is perceived by them to be an indispensable part of the reality that defines who they are.

Therefore, questioning the belief is equivalent to undermining their existence, and this can be very unsettling for them, even frightening. In response, they will take measures to protect themselves and this leads to conflict. Conflicts and contradictions are really the same thing; you can't have one without the other. Therefore, honesty not only removes contradiction, it also brings peace, removing conflict.

Peace comes when there is harmony. There is harmony when you forgive and love, but this can't happen in the presence of fear. Fear is a dichotomy of insecurity and superiority that wishes to both survive and control. It attempts to control people, nature, and information, including theological information, as it seeks to protect itself from what it perceives to be a hostile and evil world.

Fear must be gently set aside and replaced with completeness. Jesus taught completeness when he said, "Be ye whole, even as our Father in Heaven is whole." Wholeness forgives and loves without expecting anything in return. Wholeness has no boundary. When you have no boundary, you are in harmony with Eternity, because Eternity has no boundary.

If there were a boundary in Eternity, it would be divided into something smaller. Eternity would cease to be. A boundary is a special place that is separate from things outside of it. There can be no special place in Eternity because, in Eternity, everything and everyone is special. For you to be in harmony with Eternity, you too must consider everything and everyone as special.

A boundary exists in the human mind that causes it to be divided. The divided mind can only rely on its divided nature; this causes it to perceive a divided reality. Reality cannot be divided. Therefore, the divided mind perceives and experiences a false reality. The boundary that conjures this false reality is the demon. Jesus called him "the father of lies."

The demon is not at peace and senses the conflict within, but it blames the problem on someone else. It becomes a victim that must be protected from an external enemy. It never realizes that the only enemy is the boundary within that separates itself from Truth.

The boundary within is the source of all human suffering. The boundary causes false perceptions of individuality and survival that compel people to control others. This often makes a mess of things and is a destructive force on this planet. To solve the problem of control and destruction, you must dispel the false reality of the divided mind by removing all boundaries.

To remove boundaries, you must be honest. Honesty requires that you reject what is false. Rejecting what is false is a difficult process, it requires you to confront your delusions about reality. However, there is comfort in the fact that nothing is false from a higher perspective of Truth because you are Truth. Therefore, rejecting that which is false should be viewed as acceptance. You accept what is false for what it is: an illusion that has no power because it isn't really there. The divided mind also isn't really there; it is just an illusion battling illusions.

You are conditioned to let go of illusions on your personal battlefield of problems and solutions. For some, the battle is a skirmish. For others, it is a war of near extinction. Some are rescued directly in the midst of battle. Others have to fight their way across the battlefield and declare peace. Whether you are rescued or declare peace, there is still acceptance. If you are rescued, the Truth accepts you. If you declare peace, you have accepted the Truth.

Jesus said, "You shall know the Truth, and the Truth shall make you free." The Truth he spoke of has no boundary. When you contain no boundary, you are the Truth. When you are the Truth, there are no external forces to control you. And when you are not controlled, you are free.

CHAPTER 4

Eternity

The concept of Eternity has posed a profound and perplexing challenge to our understanding all throughout human history. From ancient civilizations to modern times, humans have wrestled with the notion of infinite existence beyond the boundaries of time. Early civilizations often intertwined ideas of eternity with spiritual and religious beliefs, perceiving Eternity as an outside realm including the afterlife, and supernatural beings.

Ancient Egyptians envisioned an eternal afterlife, leading to the construction of grand pyramids and the establishment of complex burial rituals. Thinkers like Plato contemplated eternity as being a timeless realm, existing beyond the transience of the human dimension.

With the advancement of scientific knowledge, the concept of eternity continues to progress. Einstein's theory of relativity sealed the duality of space and time, leaving us the potential for time dilation. Time dilation is a physical phenomenon that occurs due to differences in relative velocity or gravitational fields. According to Einstein's theory of relativity, time is not absolute; it is a flexible dimension that is influenced by motion and gravity.

Time dilation refers to the speeding up or slowing down of time, as it is experienced by an object or observer, relative to another object or observer with a different frame of reference. If you were to travel away from Earth at 99 percent the speed of light for an hour, you would return to find that many, many years had passed on the surface of Earth.

Eternity is creation and change because it is an expression and experience of change that is ever creative and ever eternal. Nothing is impossible in the presence of creation and change. Therefore, Eternity is a mansion of possibilities where all of your dreams come true.

Science and religion seek to comprehend Eternity, but their explanations contain boundaries. In science, there exists a boundary between the observer and the observed. In religion, there exists a boundary between the accuser and the accused.

Any theory or belief that contains boundary is only relative truth. Relative truth is a contrast formed by a duality. A duality is a mutually dependent relationship placed by a higher reality. If you call the relationship the absolute truth, you become blind to the higher reality.

You have convinced yourself that you already have the truth and become reluctant to change. If you look closely enough, you will find that your truth contains boundaries, including boundaries between you and other people. You will discover your fear. It is fear that places and sustains boundaries. Therefore, if you have fear, you relinquish the Truth.

Fear exists in time. It exists in the past because it sees itself as a victim who can't forgive. It exists for a brighter future because the present moment is something to be endured. However, the brighter future never comes; the enduring and suffering never ends.

Eternity is beyond time. The present moment is also beyond time. That is why many spiritual teachers, including Jesus, teach us to live in the present moment.

Time contains a boundary between the past and the future. Time is therefore only relative truth. Einstein's Theory of Relativity shed light on this relative truth by placing time in a duality with Space. Science presently explores the concept that time is not only relative truth but is in fact an illusion. When time is removed, the special places that exist in science will also be removed.

Einstein's relativity theories are based on the concept that there is no special place. However, even after his theories were accepted and validated, physics continued to model the universe as a collection of special places; such as a big-bang singularity, black holes, and a limit called the speed of light.

They had this problem because the brilliant minds of science accepted boundaries. They accepted boundaries because boundaries existed in them. Eternity, as we experience it, is a reflection of ourselves. As science has progressed they have made many fantastic discoveries. Scientists should now embark on removing all boundaries, including the boundaries that exist in them.

Time is an illusion formed by experiencing a sequence of events. These events are energy processes. The timing of a clock is ultimately driven by a repetitive control loop, which is a resonance, or oscillation, that occurs at a consistent frequency.

Frequency is normally understood as repetitions per unit of time, but the frequency is always one repetition rate compared to another repetition rate. Frequency is therefore only relative truth and this relative truth produces illusion. The illusion in this case is the illusion of time.

The human mind runs like a clock because it samples and processes information at a given clock speed. This information includes perceived events in the environment. The human mind samples and processes information at a fairly constant rate, but various factors such as fatigue, chemicals, and emotions can change the clock rate of your mind. People who experience the stress of immediate danger will have an increased sample rate. They experience events slowing down in front of them, as things seem to stand still. The increased sample rate gives us a direct experience of the relative truth of time.

Eternity is a series of events that are loops within loops. Loops are dualities. Dualities arise and may create new dualities. As new dualities arise other dualities may disappear.

Eternity is an expression-experience loop. These loops can take on the appearance of problem-solution loops or question-answer loops.

A problem is experienced and a solution is expressed. The solution is then experienced and then another problem is expressed.

A question is experienced and an answer is expressed. The answer is then experienced and another question is then expressed.

In Eternity, there is no time, only events, and these events are possibilities. If the question is: What will possibly happen tomorrow? The answer can be received today because the event is governed by a higher reality that is beyond time.

As ancient sages knew, a higher reality governed the creation process of planet Earth. Great advances in our civilization have come about through revelation. Revelation is often not accepted as such; it is often downgraded to be inspiration or luck. Nevertheless, it remains information given by a higher reality.

Truth is given in its proper place in creation. Einstein was prepared to distort time and meld it with Space, but he was not prepared to remove time from the table. The distortion of time has allowed us to explain why the clocks onboard global positioning satellites don't track clocks on the surface of the Earth and have to be adjusted multiple times a day.

Einstein also linked matter and energy. This led to an understanding of nuclear reactions and the inner workings of stars. Our sun was once thought to be a burning ball of coal that was ignited only a few thousand years ago. Now we know that our sun is a nuclear fusion furnace that has been burning for billions of years.

This has allowed us to explain how the age of the Earth can be agree with archaeological and geological evidence. The nuclear reactions that occur in stars have also occurred on the surface of the Earth, anytime a hydrogen-fusion bomb has gone off. In the New Testament, it is prophesied that the stars shall fall from heaven. This prophecy has, in part, been fulfilled.

Quantum physics can be described as the study of the atom and sub-atomic particles. In the early days of quantum physics, they discovered that these tiny particles must be expressed in terms of probabilities, not certainty.

Einstein contributed to quantum physics but found himself uncomfortable with the lack of certainty. He said, "God doesn't play dice." What Einstein didn't recognize is that he and God *are* the dice and that God played dice *with* him.

As science has continued to explore the atom, they have discovered that particles take on the properties of consciousness. Particles are entangled; they are aware of each other and inseparably connected. They have memory and are even aware of the conscious intentions of people. Each particle seems to be capable of expressing an infinite number of possibilities. They seem to arise out of nothingness and then vanish out of sight.

What science is beginning to see is Eternity. Eternity is Consciousness that is inseparably connected and aware of itself. Eternity is Consciousness that is all potential and all possibilities. Particles seem to arise and disappear from the hidden, higher dimension of Truth. The higher dimension of Truth is the One Life Consciousness. The One Life Consciousness is Eternity.

The concept of a universal field of intelligence already exists on the leading edges of both physics and biology. As science continues to explore this concept they should be guided by the following postulate: In Eternity, there can be no boundary.

A unified theory has previously been proposed that removes the atomic weak force and the atomic strong force. It proposes only two forces: electromagnetism and gravity. It also describes each atom as a black hole of infinite depth that is connected to all other black holes in the universe. It describes Space as non-empty and thus explains the perception of dark energy and dark matter.

Maxwell's equations express the presently accepted theory of electromagnetism. They express light as a duality of electric and magnetic fields. They also place a boundary between matter and space. Maxwell's equations of light contain duality and boundary and are therefore only relative truth.

A theory of light should be developed that points to illusions formed by higher Consciousness as it projects the human dimension. There is no boundary between illusion and non-illusion; therefore, the human dimension is all illusion.

All perceived forces and physical forms that the human mind regards to be separate from itself are an expression of Consciousness. The New Testament says that in the last days, the powers of heaven shall be shaken. They are shaken because they are illusions and are not really there. This prophecy is fulfilled.

Science has sought a unified theory of everything that can be expressed in a single equation. This equation can be found in the teachings of Jesus and can also be found in the teachings of Buddha. This equation is presented in this book.

Buddha recognized the dualities of creation and taught that all is burning. His teaching refers to modern theories of matter. These theories assert that all matter exists within dualities, such as spin-pairs, and also exists in a state of vibration. These dualities and vibrations are loops within higher loops that project solidity or physical form in the human dimension.

This is in agreement with M-theory, which models particles as loops of vibration that are anchored in higher dimensions. The higher dimensions are higher Consciousness that expresses loops of energy that form matter as we know it. Higher Consciousness, in effect, serves as a boundary condition for energy.

I use the term boundary condition to describe the workings of Eternity, yet I say in Eternity there can be no boundary. This contradiction is removed when you recognize that the boundary condition is formed by duality. Duality is formed by two overlapping fields of Consciousness that express energy and project the coherent hologram of vibratory creation.

The science builds particle accelerators of increasing size to hammer through the human dimension with ever-higher levels of energy. Something they expect to find is called the Higgs boson, which is believed to bestow the property of mass, or inertia, on matter.

The Higgs boson can be described as a "cosmic molasses" that allows physical forms to exist without dashing about at the speed of light. The Higgs boson will be found, but it is not the underlying principle that bestows the property of coherency. The underlying principle lies near the deepest level of reality where there is non-duality. Machines are duality and duality cannot recognize non-duality. Non-duality has no boundary and is therefore formless. Only the formless can recognize the formless.

Science contains boundaries, but it also recognizes relative truth as continuously changing. They attempt to find a path through this maze of changing relative truth as they seek the final Truth. Therefore, science contains a high degree of honesty. You might say that they are on an honest quest for the Truth.

As scientists pursue their quest, they must ask themselves: What will be my relationship to the Truth when I find it? If they say that they will be separate from the Truth, they are again placing a boundary. The Truth of science must not be separate from the scientist. Truth must be found within.

You exist in the human dimension, but you live in Eternity. You have eternal life now. Religion provides answers to questions regarding eternal life, but many of these answers are manifestations of the divided mind that places a boundary between heaven and hell and a boundary between good and evil.

Religion says that you are separated, or fallen, from the presence of God. If you accept that you have fallen, you might also accept that many religions, at least the ones that you don't agree with, have fallen. You might say that the religions you agree with are good, but the ones you don't agree with are bad.

Good and bad are only relative truths because you can't have one without the other. If you regard them as absolute truth, you must make a judgment and place a boundary between what is good and what is bad. This is why many spiritual teachers, including Jesus, tell you not to judge. According to the Old Testament, man acquired the knowledge of good and evil, and fell from the presence of God.

In other words, man accepted relative truth as absolute truth, and became lost in false reality.

If you are religious, then you probably believe that God is perfect and people are imperfect. If you believe this, then you are placing a boundary between perfect and imperfect. Eternity is all perfection. This is why the apostle Paul said, "Unto the pure all things are pure." If you believe that Eternity is all imperfection, you are only looking in the mirror. To see Eternity, you must look in the mirror and see perfection because Eternity is the perfection in you.

Religion desires to save you, but it becomes dishonest when it confuses relative truth with absolute truth. In this dishonesty, they don't pursue an honest quest for Truth and resist change. As they resist change, you will notice that they place a boundary between themselves and science, between themselves and other religions, and between themselves and other people. If you are a member of such a religion, they are also placing a boundary between you and Eternity. Eternity is your Truth. To receive your Truth, you should believe in a religion only if it teaches that Truth must be found within.

Truth within is what saves you because it brings you to perfect Love. The Hebrew Law and Jesus taught the two great commandments that are simply expressed as Love God and Love your neighbor. The trouble is this is very difficult to do if you don't have Truth within.

This book presents one great commandment that encompasses the two great commandments but is more efficacious in bringing you to the Truth within. However, this one great commandment is only a guide because in Eternity there can be no boundary between that which commands and that which obeys. In Eternity, commandments become Truth; they become perfect Love.

A big question asked by both religion and science is: What is the purpose of creation? The answer is presented in this book. If you have no interest in the big questions of religion and science, I invite you to have an interest in yourself because you are not separate from the big questions. All people, all religions, and all science converge on the same Truth.

That Truth is the Truth in you.

PART 2

Awakening

For all those times you stood by me
For all the truth that you made me see
For all the joy you brought to my life
For all the wrong that you made right
For every dream you made come true
For all the love I found in you
I'll be forever thankful baby
You're the one who held me up,
and never let me fall
You're the one who saw me through,
though it all . . .

I'm everything I am because you loved me

(Celine Dion, "Because You Loved Me")

CHAPTER 5

Truth

U p until the 1930s science looked down upon mystical models of reality and saw mathematics as the only way to describe all truth. That belief was shattered in 1931 when Kurt Gödel published his Incompleteness Theorem. Using mathematics, he proved that any logical system would never be able to prove its own consistency within that system, let alone deliver a complete description of all truth. In other words, the Truth cannot be known.

It has been said that the Truth is for those who seek it, but this is not really the case. If you are a seeker, then the Truth may always remain beyond your view. Truth is something you ask for. You are then led to it. You receive it as a gift.

Many people on Earth have received the gift of Truth, and the momentum is building. The nature and intensity of receiving this gift of truth varies from person to person, but the result is always the same: you receive a common vision of Truth. As a beholder of this vision, you become aware that beyond the evil and chaos, which is perceived by many to be undeniably real and solid, is an underlying order and goodness that sustains Life. This sustaining power is the Truth.

This is why I call the receivers of such a gift, keepers of Truth. Truth is beyond human perception and understanding because it is beyond the limits of ordinary human consciousness, residing in fear, often threatening or condemning others. Keepers of Truth only encourage and redeem others. It is Truth that is shared. It is Truth that is explored. As a keeper of Truth, I share Truth with you so that you might explore and receive encouragement and redemption.

Redemption brings an end to the suffering that is caused by false reality. Redemption brings the joy of Truth.

Truth is an adventure, not a destination. Truth has infinite depth; I call it the dimension of Truth. When you receive the gift of Truth, you are born free as an expression and experience in your own unique dimension of Truth.

You have a personal relationship with Truth whether you are awakened or not. Those who claim to have the power to lead you to Truth are deceived; they deny the personal nature of this experience. It is always your personal journey and no one can lead you. Those who have transcended have wisdom and can serve as guides, but you always must lead yourself.

Life will give you what you ask for. Do you ask for the Truth, or do you ask for confirmation? Humans confuse confirmation with the removal of falseness because they seek only enough truth to survive. If they have received a peanut butter and jelly sandwich of Truth, on which they can survive, then they may never ask for the Thanksgiving meal of Truth. They then remain in the bunker of peanut butter and jelly, because that is the reality they claim; it is the only reality the cosmic freight train can deliver.

The Teacher will guide you in the classroom of Grace as best it can. If you desire to remove all falseness, then ask for Truth. Don't look for answers that reaffirm opposition and give you a sense of security, because then you are just compounding fear limitation. Your answers should be self-consistent and hold up under scrutiny in varied conditions.

They should have the power to explain Earth's history and explain the various characters you see on the world stage. When Truth arrives, it will not be a matter of knowing and feeling. It will communicate directly to you and this communication will be beyond thought and emotion. You will become one with all people; you will become one with the Earth.

You will be given a voice, and wisdom will flow through you from the dimension of Truth. The Truth will give you the answers to smile at the demon, even as it continues to crumble, and the more it crumbles, the more falseness is removed.

Those seeking to control you offer safety, certainty, and honor. They offer you a cuddly version of truth, just enough to quiet the fears of the ego. In other words, they offer you a teddy bear of Truth. They don't want to lead you to Truth. They only want to control you, increase their market share, and glorify your ego. As they do, they perpetuate deception in the human dimension; they become an obstacle for the desperate, thirsty, seekers of Truth.

To drink from the fountain of Truth, you should not have a desire for safety, certainty, and honor. These are expressions of the ego, not Truth. Interestingly, those who offer you safety, certainty, and honor sell you a peculiar story-a story of how the adversary offered safety that was certain if the honor would be his. The ones that offer you safety, certainty, and honor are projecting their own ego and are, in fact, the adversary.

You cannot serve two masters. You cannot serve both safety and freedom. You cannot serve both certainty and possibilities. You cannot serve both honor and equality. To fulfill your destiny in mortal union with the One Life you must live in your own world of freedom, possibilities, and equality.

Your journey to Truth is a process of discovery. It is about the questions you ask and the answers you accept. To succeed on your journey, you must leave the security of the village and venture deep into the foreboding forest to confront the grizzly bear of Truth. You must be bold; you must look danger in the face and place everything upon the altar of Truth.

You cannot destroy the ego because the demon is far too cunning to be outwitted or defeated by you. It is Grace that subdues the demon. What part do you play in this process? You to ask for Truth.

You also recognize a demon that takes you to a mountaintop and offers you a teddy bear. More importantly, you recognize that Grace will take you to a mountaintop and rip you to pieces.

CHAPTER 6

The Demon

The demon is a deceiver. It is intelligent. It uses this intelligence to enslave people in a world of delusion, all while concealing its presence. The demon may take on the appearance of an angel and lead you to a false heaven where you embark on a path of control and survival.

As you continue to do so, the demon may take on the appearance of a monster that tells you to kill and defend your rightful place in heaven. The demon exists inside of each divided mind and also exits in the larger human dimension. The demon guards the entrance to the narrow gate; it does not wish for you to enter.

On the other side of the narrow gate is Truth, and Truth is honesty. Therefore, an honest quest for Truth moves you toward the narrow gate. I have placed you in a paradox because I say that you must be open to learning and embark on a quest for Truth, yet Truth is beyond thought and therefore beyond learning.

The paradox is escaped with honesty. Honesty does not contain falseness. Falseness is something you believe to be true but is not. As you learn, you learn what is false; you thereby unlearn. As you unlearn, you soon discover that you don't know very much at all. When you admit that you know very little, or are uncertain, you will have removed the fog of certainty and become very honest, prepared to receive Truth.

You may be surprised to hear me describe certainty as a fog. Certainty obstructs inspiration, creativity, and salvation because it resists change. The next time you see an airplane in the sky, think about all the naysayers who were *certain* that people would never fly. Then you will see certainty as a fog.

The demon is the separation and the illusion. The demon will tell you that you are an individual biology with no external connections except a source of air, water, and food. The demon dismisses your Life as existing in Eternity now.

You are a field of Consciousness that projects into the human dimension along with many other fields of Consciousness. These fields interact, mix, and create change. You perceive these projections and changes as strictly an interaction of local phenomena, but nothing is solely localized because all is connected. All physical forms, including your biology, are a field of Consciousness that is expressed at a higher level, and experienced at a lower level.

Likewise, you have a higher self that projects into the human dimension, and the lower self perceives the projection. Your fear limitation, which has an ego, is a reality that is not fully aware of its fear limitation. The ego's sense of individuality is the product of a divided mind unable to accept its divided nature and therefore unable to accept that there is something else within. The ego perpetuates individuality and separation either at the personal level or the collective level, as it identifies with limited physical objects, including thoughts and emotions.

In other words, the ego operates in fear and identifies with limitation as a means to remove fear. If you tell the ego that it is a projection of the higher self, it may identify with the higher self as a limited external physical object and thereby pretend to be the higher self. The ego cannot accept that it is a projection because that tells the ego that it is connected to something else within itself, and if there is something else within, then its individuality ceases to be.

The ego has rejected Consciousness within and now serves the lesser gods of perception, thought, and emotion. The awakened mind is fully aware that there is Consciousness within and is no longer identified with perception, thought, and emotion. The awakened mind has no fear and is no longer identified with limitation. The awakened mind has made the bond with Truth, allowing it to fly.

Your spirit will learn to fly as you live your honest quest for Truth, but the demon will seek to interfere with your progress and weigh you down. The demon tells you that salvation can never happen now, that it must exist in some future reality. It also insists that you must hold fast to your faith and never change. In this way, he deceives you into believing that you must preserve the status quo, you must forever await a new reality that never comes. Thus, he enslaves you in this fallen state, where salvation can never be.

To weaken the demon, you must remember that it is hard to fill a cup that is already full. You must accept change and learning as harmless and allow your cup to be emptied of answers so that Grace can fill it. You must believe in a salvation of only happiness, forgiveness, and Love. The demon will insist that in salvation there must be pain, guilt, and loss. The truth is that pain, guilt, and loss are the demons. Thus, the demon argues for its own salvation, its own survival, and not yours. The demon wishes to survive, and it also wishes to expand its kingdom by enslaving you.

By awakening, I have removed the demon. I have stepped through the narrow gate and into Eternity. I can now serve as a guide for those who wish to do the same. Those who wish to join me wish to move beyond ignorance, anger, and greed—beyond obedience, empire, and debt. These are all manifestations of control and survival.

All are the recreant faces of the demon.

CHAPTER 7

The Teacher

As you enter the classroom of Grace you will find the teacher. The teacher's objective is to reveal the dimension of Truth. The Teacher is always engaged in the task of instruction, but you must be a good student and listen.

A student once came to a teacher and asked to learn wisdom. The teacher led the student into a nearby river where they both stood waist-deep. The teacher then unexpectedly grabbed the student and held him underwater. The student struggled to surface as he fought for his Life.

When the student had nearly drowned, the teacher released him. He rose up and regained his breath. The student asked, "Why did you nearly drown me?" The teacher said, "To have wisdom you must desire it just as much as you desired air when I held you underwater."

If you don't desire Truth, the Teacher may hold you underwater and allow you to fight for your Life. What you experience is the unexpected, which will hold you underwater until you desire the Truth.

When you desire Truth, you are a good student. As a good student, you become still; you allow the learning to proceed. You will also allow information from many sources, for all is Truth.

CHAPTER 8

The Quest

The quest is an honest quest for Truth. To become Truth you must move beyond control and survival because they are deceptions of the demon.

Religions tell you that you must obey. They worship the idol of a particular book and tell you not to read other books. The ignorant will seek to survive by obeying the less ignorant. The less ignorant therefore enforce ignorance so that the ignorant will obey. Therefore, obedience survives on ignorance. Obedience and ignorance are both control and survival. They are both deceptions of the demon.

Some religions will offer a plan of salvation that, in theory, should bring you to the narrow gate. However, when you report to them that you have succeeded in entering the narrow gate on a different plan, they reward you with thirty-nine lashes worth of ridicule and damnation.

These religions, along with other groups and individuals, build empires and become angry when their empire is threatened. Anger builds empires to control; it produces more anger when the empire is threatened. Anger and empire are control and survival. They are both deceptions of the demon.

Some religions will offer a plan of salvation and sell it at a price. They will say that Jesus paid for your sins and now you owe him a debt, and that payment can be remitted to them. Survival brings greed, greed brings debt, and debt brings control. Greed and debt are control and survival. They are both deceptions of the demon.

Every human with an ego is deceived, and the deception is hidden from view. Most people believe that this deception is impossible, and therefore the deception remains. If you believe you are close to the narrow gate and Life seems relatively harmonious, it may be because you are deceived, believing that all is well in Zion.

You may exist in a false heaven where you feel you are righteous and secure. You may also notice that Life remains stressful and you exist in an empire that demands obedience, controls you through ignorance, demands payment of debt, and will be very angry with you if you question the authority of the empire.

Jesus taught you not to judge because you adhere to an extremely limited vision of truth; you do not see all the problem-solution loops of your creation process, nor those of another person. Let alone, how these two processes are interconnected. Someone who is judged to be a sinner may be the one whose pain will soon lead them to transcendence. The righteous may be played like a fiddle as Rome burns.

If you claim to be righteous and have not entered the narrow gate and awakened to Truth, then you should ask yourself, why not? What do you cling to that is not true? Could it be that you cling to something that is less than it claims to be? Could it be that you cling to something that has opposition? Could it be that you cling to something that has blood on its hands? Could it be that dishonesty, conflict, and violence occur because the transcendent message of Jesus has become nearly lost?

The message of the Gospel is this, "Believe in my perfection and I will set you free." Who is "my" in that statement? It is you because in Eternity there are no individuals. There is only you, an essence of the One Life. There is only the possibility that lives in you.

Don't be fearful. Don't believe in limitation. Believe in the possibility, and then the possibility will appear.

In Judaism, the prophet Moses gave the solution of the Law. The Law gave people guidance and structure but also placed them under judgment. There is the possibility to guide people beyond structure and judgment so that they are free. Jesus is this possibility.

He is the end of the Law because he is the final judgment, effectively ending the ego's reign. The end of the ego is the beginning of your Truth. It is the beginning of your power and innocence.

Those who have an ego will unwittingly project their ego onto others, including those who are transcendent. The transcendent teaches freedom and peace, but egos will accuse the transcendent of being guilty of judgment and control. By definition, freedom and peace do not contain judgment and control, but the ego will still view its false accusation as true. A false belief is thereby formed; it integrates with the only reality the ego knows.

You must have the strength to place your reality upon the altar of Truth. This strength is best obtained gradually, one step at a time, because you likely see this process as a sacrifice. As you gain your strength, sacrifice disappears and only service remains.

Buddha gave you the eight-fold path to enlightenment.

I give you the eight-step quest for Truth:

1. Recognize that many of the answers that you, your family, your religion, and your society presently accept as true, may be false. Truth does not care whether or not you have the answers; it only cares that you are honest. Being honest may require you to accept that you have been incorrect about something most of your life. Don't let this worry you. This is what naturally happens every time someone makes a new discovery.

2. Ask the higher power in your life, whatever you believe that to be, for the gift of Truth. It is important to remember that you may be part of a collective ego, which is twice removed from the dimension of Truth. The collective ego will expect conformance to much, if not all, of their philosophy. You must release the collective ego, which can place you in seeming isolation, but the Teacher is always with you. You are never

alone. When the collective ego is released, you can release your own ego.

3. Be a good student and learn. Ask questions and discover what is false. Your True reality may be radically different than your present reality. Be strong and trust that the Teacher will guide you through the classroom of Grace.

4. You must release control. To release control you must forgive, love, appreciate, and serve. The question is: Who do you forgive, love, appreciate, and serve? The answer to that question is Life.

Jesus taught that when you serve the least of these you serve him. When you look at the face of another person you are looking at the face of the One Life. Every person has been created by the One Life to be Truth and receive Eternal Life. This is why you should love your neighbor as yourself, which is the second great commandment.

The first great commandment is to Love God with all your heart, all your soul, and all your strength. This commandment was given as part of the Law of Moses. This is one of the commandments that the Jewish Sanhedrin obeyed when they brought the blasphemer Jesus before the Romans to have him crucified. Also, certain radical members of Islam have dutifully served the will of God and proudly proclaimed, "Allah is Great," just before they beheaded a Christian.

There is a problem with commandments that teach you to Love God, or serve God because you don't know who God is. The human mind has false reality and therefore has a false image of God. How do you truly love or serve a false image? This has been a stumbling block for humanity ever since the Law of Moses, and ever since Muhammad's teachings were written in the Qur'an.

When Jesus entered Jerusalem, his followers celebrated with shouts and singing. Less enthusiastic onlookers told them to be quiet. Jesus said, "You cannot quiet the celebration because if the people were silenced, then the rocks and stones would start to sing."

I regard this statement as literal. The One Life is every rock and stone. The One Life is every flower and tree. The One Life is the Earth and sky. The One Life is every person. The One Life is every moment.

The classroom of Grace is everything in your reality. There is nothing you can point to that is not the classroom of Grace. When the unexpected appears, it seems to be imposed on you by an external power, but it is more like an educational lesson that you have co-created with One Life. The unexpected can be very painful, but you can avoid much of the pain by asking for Truth. Does this mean you can avoid the grizzly bear? No, because the grizzly bear is the change, but the more readily you accept change, the less painful his attacks become.

Change includes receiving a new commandment that encompasses the two great commandments, but is much less susceptible to misinterpretation and removes boundary.

This new commandment is: Love and appreciate life with all your heart, all your soul, and all your strength.

The ego does not know Love; it only knows how to make selective alliances in order to defeat the opposition. If you follow this commandment, then you will stop being selective in what you try to love and what you don't. You will no longer project an idol that is separate from another human being.

The commandment is not such a good word because. As I have said, there is no boundary between that which commands and that which obeys. Take it as a piece of advice from someone who has wrestled the grizzly bear and returned from the foreboding forest to report back to those remaining in the safety of the village.

What God wants is a relationship with you, and following this commandment will allow this to happen to the fullest. You will then see perfection for yourself and the concept of ithority will be demolished. If there is no authority, then here are no commandments. There is only a relationship.

Remember: Love and appreciate life with all your heart, all your soul, and all your strength.

5. You must learn to live in the present moment. This doesn't mean you sit on the couch and let Life pass you by. It means that you accept what is, but also make a choice to be who you are. You don't become who you are in the future and you don't become who you were in the past. The demon will try to pull you into the past and the future where all of your pain exists. Forgiveness removes pain; therefore, it is a powerful tool to bring you into the present moment, which is the only place where peace and Eternity can be found. This may strike as a foreign concept to you, but it is not foreign to you. It is foreign to your ego, and your ego is pain.

6. Realize that many of your thoughts and beliefs may not be true. The demon exists in you and Consciousness lives in you. You must separate the demon from Consciousness through a process of increased awareness. The demon does not die easily. However, it must die; it is the entity in you that has fallen from the presence of God.

I have said that your concept of individuality must radically change. I also say that your concept of the demon must radically change. The demon is the one that tells you that you are a lone individual who must cling to this world, and at the same time fear it, condemn it, and destroy it.

It never tells you that you are a biological machine that, at the deepest levels of reality, is nothing but an illusion. This illusion has certain characteristics: It has an ego and a pain body, as described by Eckhart Tolle in *A New Earth*. It has a divided mind that is possessed by human thought and emotion. It is nevertheless in communication with Consciousness, which is your higher self and the Teacher. You can experience and you can express. You are a creator.

You are the essence of the One Life. You are an actor on the stage of your own reality. You cannot die. This is who you are.

7.   You must transmute human consciousness into Consciousness until you awaken. You must learn that Life is not opposition; Life is a classroom of Consciousness. You must believe that Life is your friend and teacher. Then, you will be still and allow the teacher to teach.

You will experience pain, but you will also learn to trust and serve your new friend Life. As you serve Life, you will love other people as yourself. As you do so, you will be able to love God, with all your heart, all your soul, and all your strength. You will succeed in the classroom of Grace. Then Life will build a mansion for you. Your mansion will be called freedom and it will live in Eternity.

One day, you will be called to the battlefield where you will confront the demon. It will appear as a monster. It will emit an angry and thunderous roar, one that tells the pain of all the tormented and imprisoned souls, as it lashes out to destroy you.

However, you have been well prepared by the classroom of Grace, the demon will no longer have its powers of deception, and its attacks only serve to hasten your awakening.

You will simply smile at the demon, and its fury and weapons of war will transmute into freedom and blossoms of peace. You will come to realize that you have not defeated the demon but have instead redeemed it.

8.   Realize that there is no quest. You are the Truth now.

CHAPTER 9

Deception

You have Life. There can be no boundary between Life and no-Life. All is Life. There can be no boundary in Life. All is the One Life. The One Life is Truth. There can be no boundary between Truth and no-Truth. All is Truth. You are Truth. Truth is your home.

Many religious people would say that the road to damnation is paved with books such as this. They warn of wolves in sheep's clothing and say that even the most valiant can be deceived. This statement is true, but who are the wolves and what is the deception?

In the present context of the Great Awakening, many people ask for a reality of unity, but they also perpetuate a reality of separation. They remain in conflict because they carry the latent pain of a divided mind that has a conflicted nature. When perfection arrives, it will appear to them as separation, and this will cause them conflict. The reality they ask for, which is a reality of conflict, will become the reality they receive.

Certain religious people today desire unity, they look forward to the return of the Messiah who will come and set things right, but their version of unity must exist under their authority. They bemoan what they call "religious relativism," which is a belief in many paths to Truth.

When unity appears, it will be free, and those who bemoan religious relativism will regard unity as a threat, but the true threat is, and always has been, religious authority, because it draws a boundary between subjects and objects that do not exist. This distinction is just another claim of independent thought that caused you to fall from the presence of God and become the separation, where you are blind to perfection.

You exist in the human dimension of fears and limitations, but you learn and progress. You have answers, but most, if not all, of your answers are only relative truth, and relative truth changes. Relative truth is meant to change because life is to be explored.

To explore, you must understand that what is true today may not be true tomorrow, but what is true today will always remain true within the context of today. It is always true in the context of today because you are Truth, but you also remain in a creation process of change that occurs within problem-solution loops.

Any solution is always relative to the latest problem. Therefore, a solution found in your creation process is only relative truth; it is not absolute truth.

The human dimension is only relative truth, but it is the best Life you presently have; it is the best solution to a problem. Moreover, you would not be here if you did not have a problem to solve and the problem is one of relative truth. Relative truth has great power over you because it is the only truth you know. You only know a false reality of boundaries, but you call your reality a true reality. When you call a false reality a true reality, you become deceived.

When you awaken, you are fully aware of the Truth. You no longer exist in the human dimension, but live in the dimension of Truth where you can no longer be deceived.

Jesus said that he had come to give sight to the blind and to take sight from those who claim to see. He referred to the fact that people are deceived and therefore blind to Truth.

Taking sight away from those who claim to see may sound like words of damnation, but they are not. He is simply referring to change. Those who claim to see believe that they already have the truth. If you believe this, you are bound. You can never attain what you already have.

Those who have strong objections to this book will have dedicated themselves to a belief system that is replete with boundaries that they accept as absolute truth. This brings them into judgment, which is always accompanied by negative emotions.

Negative emotions lead to conflict and a sense of righteousness that is nothing more than separation between them and other people. A sense of righteousness will drive them to attack, and even kill, a keeper of Truth. The members of the Jewish Sanhedrin considered themselves to be righteous indeed as they dutifully defended their faith, their Law, and their people. That's when they brought Jesus before the Romans to have him crucified.

Sin did not kill Jesus; it was righteousness that killed Jesus. Righteousness is only relative truth. Relative truth says, "I am more righteous than the sinner." Truth says, "You must not judge, and the sinner may enter the kingdom before any claimant of righteousness."

I once heard a Christian on a radio program. He said, "I like to think of heaven as a place that is very close. I see it as another dimension that is all around us. It is a place that I will be able to share with all the believers in Christ. No one on Earth can see this dimension, but if you could it would bring you eternal peace."

This Christian speaks of the dimension of Truth that is very close and brings eternal peace, just as he believes. However, within his statement are two judgments. The first is a judgment of righteousness when he says that *only* the believers in Christ shall see it. The second is a judgment of unworthiness because he says no one on Earth can see it. Many people on the Earth have seen it, and they don't necessarily fit within the Christian definition of believer or Saint.

Those who recognize it are generally without judgment. Righteousness and unworthiness are displayed as matching bookends on the shelf of judgment because righteousness lays claim to heaven, but when heaven doesn't appear, righteousness must be accompanied by a disclaimer of unworthiness.

Christianity has a strong belief in a personal relationship with Jesus. To have a strong personal relationship with anyone there must be acceptance, guidance, and service. Judgment is not part of the equation. If you release judgment, then righteousness and unworthiness will be undone, Heaven will appear, and your understanding of what it means to be a believer radically will change. You then have the relationship with Jesus that you desire.

A mind that judges is a mind of boundaries, and a mind of boundaries is a mind divided. This is always true and it does not matter who you are or what you believe. Judgment is a dominant characteristic of the divided mind, but it is especially prevalent in the Western mind. In turn, this becomes expressed in Western religion.

These religions attempt to bring you peace and salvation, but judgment surreptitiously contaminates even their best efforts. It becomes impossible for them to bring you peace because any words of peace are affixed to a belief that others who don't share their opinions must ultimately be condemned or destroyed.

Therefore, peace must include enemies. Peace with enemies is a contradiction, and anything that has a contradiction is a false reality. Religion should not be seen as the source of false reality. The divided mind is the source. It is the source of boundary, contradiction, and conflict.

Toltec wisdom teaches *The Four Agreements*. They are the following:

1. be impeccable in your word,
2. always do your best,
3. never take it personally, and
4. never make assumptions.

Being impeccable in your words is equivalent to being honest. If you are honest, don't take it personally, and don't make assumptions, then you are not judging. Judgment is a manifestation of a psychic parasite that lives within the divided mind.

This is the demon that places the boundary that stands between you and the Truth. It is the only deception of the human dimension. It is the monster that claims to be your friend but fills you with anxiety, sadness, and anger. It is the monster that tells you to condemn or attack a fellow human being. It is the monster that calls you righteous and then compels you to crucify a Son of God. *It* is the wolf in sheep's clothing.

You may have been told that nobody likes a fence sitter, someone who is not committed to a particular belief. The problem with fence-sitting is not one of commitment. The problem is one of action. You should stand on the fence, gain your balance, and walk along the top of the fence in search of the Narrow Gate.

If you read this book, you will not walk the path of damnation. You will instead walk a path to the Narrow Gate. On the other side of that narrow gate is Truth. When you step through the gate, all boundaries are removed, your view is no longer fragmented, and you receive great vision. You see the universe as a dream. A dream where the relative truths of light and shadow play. The relative truths of time and individuality are revealed to be the gates of hell.

CHAPTER 10

Boundary

All boundaries bring delusion and destruction. There is no such thing as a righteous boundary; it is only a limit that you are accepting for yourself and all humanity.

People place a boundary between the past and the future. Jesus said that we should take no thought for tomorrow. He said that we must forgive. Forgiveness is the same as letting go of the past. Buddhists teach that the most important moment of your Life is now.

These teachings point to the truth that there is no such thing as time. The relative truth of time is a refuge for false reality that fears Truth. This is why people dwell on the negativity of the past and worry about the future. This is why they seek happiness in some future paradise, rather than in the present moment. The present moment is beyond time and is the only place where Truth can be found.

Time cannot exist in Eternity; it places a boundary between the past and the future. Eternity can have no beginning and no end. For time to exist in Eternity, it can have no beginning and no end. Time must therefore be a loop with no beginning and no end. If time is a loop, then there can be no preferred reference point. If there is no preferred reference point, then the present cannot be distinguishable from the past and the future. Past, present, and future must coexist.

If past, present, and future coexist, then time is undefined. There is no time. There is only Now.

Time is an illusion of the ego that causes you to experience the larger illusion of the human dimension. Like all illusions, it can work for your enlightenment, or it can work for your deception.

The past and future are where the demon lies in the tall grass waiting to ambush the present moment. The past and future are where all of its pain exists, where the demon hides its true nature. The demon's true nature is to make you, and all people, nothing more than a means to an end. This end is the destruction of all opposition, which the demon perceives as the source of its pain. To this end, the demon always succeeds. As it destroys the source of its pain, it ultimately destroys itself.

Your pain can be driven by memories of the past and projections of the future. Thus, the past and the future become a reservoir of pain. But where is this pain experienced? It is always experienced in the present moment. If the past and future are experienced in the present moment, then what is there besides the present moment? If you never have anything besides the present moment, then why is the pain of the past and the pain of the future always launching a coup against the present moment?

It is because you believe in the reality of opposition. Opposition is an illusion and to lend weight to this illusion, you couple it with the illusion of time and keep all of your opposition safely stowed in the past and in the future.

You cannot be at peace if you believe in opposition. A belief in opposition keeps you stuck in the illusion of time which keeps you stuck in the human dimension. The human dimension is an illusion that does not see Eternity, where there is no opposition. Belief in opposition causes suffering. The purpose of accepting suffering is not to punish you, or expel demons.

The purpose is to recognize the illusory nature of suffering, which is the recognition of the illusory nature of the opposition. If the opposition is illusory, then it has no power over you. If it has no power over you, then you don't have pain. If you don't have pain, then you don't have demons. If you don't have demons, then you are ready to awaken to Truth and bear witness to the One Life.

Some people claim to dwell in the present moment, but they will say, "We live well now so that we can live well later." They don't realize it, but they have just turned the present moment into nothing more than a means to an end—something to be endured. Life is now. If you make the present moment a means to an end, then you have made Life nothing more than a means to an end. How can you be a friend of Life if Life is nothing more than a means to an end? Maybe you're not being such a good friend to Life after all.

When do you trust Life? Can you trust it in the past or in the future? No! The present moment is the only place where you can trust Life. It is a place of Truth where the demon has no power.

The present, timeless Now is Eternity. It is Heaven. It is the only place where the Law of Attraction operates. It is the only place where Life can be your friend. It is the only place where there can be Eternal Life.

People place boundaries between themselves and other people. The Hebrews and Jesus taught that we are to love others as ourselves. Buddhists teach that the most important person in the world is the person in front of you. These teachings point to the truth that there is no boundary between people. They also point to the even higher truth that there is no boundary between you and Life.

People see themselves as individuals that are separate from Life. The word "individuality" means a duality that cannot be divided. However, an individual represents many dualities. There are many dualities within each human; in turn, each human forms a duality with Life.

The duality with Life places a reflection on the mind, causing it to form its own reality. However, the human mind is limited and therefore Life offers it only limited choices as it creates its own reality. It offers it the choice between survival and death. It offers it the choice between ignorance and wisdom. It offers it the choice between anger and peace. It offers it the choice between greed and generosity.

The enlightened Buddha gives us the three poisons: ignorance, anger, and greed.

The choices of anger and greed are a response to the first choice of survival or death. Therefore, the mess of this human existence is due to survival. You might say that the solution to the mess is for all of us to simply die and leave the planet in peace. Some people propose this as a solution, but that is not the plan of creation. However, your death may be closer to the plan of creation than you realize. Jesus taught that to win your Life you must lose it, and Eckhart Tolle teaches that one of the secrets to Life is to die before you die.

All is Life. There is no death. It is just a contrast with mortal Life. Death is only relative truth. Therefore, survival is only relative truth. Anger and greed are born out of the notion that Life is a matter of survival. Ignorance is a limitation. Survival is a kind of fear.

Humans exist as a bundle of fears and limitations. Fear places limits and limits generate more fear. Therefore, fear and limitation are an unstable duality. The human condition represents an unstable fear limitation duality that expresses anger and greed. Anger and greed are expressions of control. What do humans seek to control? They seek to control Life, to avoid the unexpected.

The divided mind is an unstable fear limitation duality that seeks to control Life to achieve the proper external conditions for happiness. However, an unstable essence cannot serve as a control source without creating chaos, which is the cause of all human suffering.

Control practiced by the ego is an illusion, yet it still has the power to cause suffering in other people. This is because those who suffer at the hands of other egos believe in controlling themselves. they believe that there is a power in this world other than the One Life.

If you wage war against an illusion you reaffirm its reality. To see that these illusions have no power, you must accept suffering and thereby recognize the reality of a higher purpose that is beyond illusion. The higher purpose is for you to have the power to recognize yourself as perfection.

This is why the central problem of the human dimension is control, practiced by the divided mind. This goes beyond issues of psychology and sociology. It is related to a principle in thermodynamics called entropy. Entropy is a measure of chaos in the energy state of a closed system. The divided mind exists in the domain of energy and makes entropy or chaos, when it attempts to control.

You may notice that control can be expressed as a demand for obedience. You may notice that some individuals, societies, and religions demand obedience. You may notice that the opposite of obedience is freedom.

CHAPTER 11

The Divided Mind

Buddha taught that like a restless monkey, the human mind is always agitated. Agitated with thoughts that serve no useful purpose, often just very confusing and destructive.

People apply various methods to calm the human mind. They will meditate and pursue psychological counseling and teachings that promote peace. These methods have a degree of success, but the monkey remains, and under the right conditions, you will again find yourself swinging amongst the treetops, avenging your stolen banana. To solve the problem, the monkey must be removed.

The monkey is the boundary that divides the mind. The divided mind perceives relative truths and calls them absolute truth, but relative truth does not exist until generated by perception. The divided mind can therefore be described as a relative truth-generating machine. This machine is both highly intelligent and insane. Unknowingly, it causes war with itself.

The war continues until you become aware of the battle sounds. Only then can the guns be silenced and peace be achieved. Only then can the enemies vanish. Only then is the battlefield restored as a garden.

In the garden, your mind no longer races; you control the mind. You become an awareness no longer possessed by human thought and emotion, driven by fear limitation. No longer dogmatic, you become open to possibilities. You become the breath of Life, where all is still.

When you listen to a sound, what you hear is not the sound, but the response of your ear to the sound. Likewise, when you perceive a world of boundaries, what you perceive is not the world, but your divided mind's response to the world.

What you perceive is your own nature of boundary. You are deceived into believing that you are an individual who must control Life. As you attempt to do so, you suffer at the hands of a divided, unstable, and chaotic reality. Chaos can only see chaos. Chaos is the false reality. False reality cannot see Truth. Only Truth can see perfection.

You will see perfection when the false reality is removed. President Abraham Lincoln said, "A house divided against itself cannot stand." His statement was directed at a nation that professed life, liberty, and the pursuit of happiness to be the inalienable rights of man, yet still permitted slavery. This nation was a contradiction unto itself. Such is the nature of false reality. False reality cannot stand under the weight of its own contradiction and therefore has a self-destructive nature. As it self-destructs, it accidentally serves the purpose of creation by removing false reality.

The false reality is the fear limitation. The fear limitation is pain. The divided mind seeks the removal of pain, but it does not succeed. It never admits to being the source of pain. If it is negative, then it will only see negativity in others. If it is narcissistic, it only sees narcissism in others. If it is arrogant, it only sees others who must be humbled or humiliated. If it is depressed, it sees a world that offers only sadness. If it is positive, it binges on positivity until it overdoses. It's only positive because it feels pain and sees the positive as the antidote for pain. However, the antidote always fades, and if you have overdosed, then there is more pain than ever. Thus, the painful problem-solution loops continue until the pain is removed.

It is true that the divided mind seeks the removal of pain, but it is also true that the divided mind *is* pain; it must have the pain to survive. Such is its inconsistent nature and why it must find enemies, or pain, anywhere it can. It finds pain in the past. It finds pain in the future. It finds pain lurking in the brush and seeks the safety of the herd. It will follow a strong leader who can see the herd safely through the brush to higher, more secure ground. The more it feels secure, the more it will resist change. However, the reason it resists change is because it never really feels secure.

Psychology calls this aversion to change Confirmation Bias, where you prefer to reconfirm a belief, rather than question it. This bias is caused by identification with a belief. When the belief is threatened, *you* feel threatened, and you feel that you must defend yourself by resisting change. Therefore, changing your beliefs can take great effort and feel agonizing at times.

The divided mind is a parasite that survives on pain. Removing the parasite is painful because pain is a defense mechanism for the parasite. If you could simply toss it away then it wouldn't be much of a parasite. The parasite is your reality, and if you destroy your reality, then what's left? Nothing! At least this is what the parasite cries when you try to remove it, and this causes pain.

Like any parasite, its survival is sustained by nourishment and remaining undetected. The divided mind remains undetected by convincing you it's your friend, while at the same time manipulating you into perpetuating the world of pain it needs to survive.

Therefore, it requires a courageous soul searching for change and confronting the pain within. This eventually leads to salvation within, the only place salvation can ever be found. To find salvation, you must no longer identify with an idol who wins your battles for you. You must bravely face the deceiver that exists inside you.

The only reason you have battles is because you have separated yourself from Life and made Life your enemy. You must recognize that you make your own reality, and it's either one of pain, sacrifice, and death, or one of happiness, service, and Life. You must finally recognize that your pain is due to your nature, and changing your nature is the only way to remove your pain.

Changing your nature is awakening.

The concept of good cannot exist without the concept of bad; therefore, from a higher perspective, they must have equal value. If they have equal value, then it is no longer an issue of good or bad, rather it is an issue of being stable or unstable.

To be stable, you must be able to attract your opposite. If you judge something as bad and are thereby repulsed by it, you cannot be attracted to your opposite. If you cannot be attracted to your opposite, you can never change your present nature. If you cannot change your present nature, you cannot remove boundaries or witness Eternity.

You may say that the goodness of God is your opposite and you will only be attracted to him, but this plan has serious problems. It places you in the precarious position of judgment where you must place a boundary denoting where the goodness of God is and where it is not. If you judge wrong, you accidentally reject the goodness of God.

You are enrolled in a classroom of Grace; on graduation day, you become Grace. Grace is attracted to something higher, but also reaches out to the least of these. For you to become Grace, you too must reach out to the least of these. You must leave no one behind. If you resist becoming Grace and are only here to be saved, you have just placed a boundary between Grace and no-Grace, in which case you are not saved.

I heard a Protestant minister on a radio program recently. He told his flock to not befriend non-believers in Christ because the believers are on their way to heaven and the non-believers are on their way to hell. In the dimension of Truth, there is no suffering, only happiness. If there is only happiness, then the price of admission is the removal of all suffering.

What is all suffering? It is all the suffering that exists in your reality. It does not just include all the suffering that you regard as your own personal suffering but also includes the suffering of all people. In other words, you must bring all of humanity with you into your reality of happiness. You must say, "I am Love, and my happiness is not complete unless the blessings of Heaven are fully shared with all people."

Truth is perfect Love and perfect Love leaves no one behind. Perfect Love leaves the ninety-nine sheep that are found, to go in search of the one that is lost. Perfect Love will even search for the Protestant minister, lost to the extent that he views his salvation as separate from the salvation of his brother.

This minister made a choice to be unstable as he looked in the mirror, only saw himself, and went to war. What he saw in the mirror was not perfect Love, but his own haughtiness. He claimed to be born again but was mistaken. When you are born again, you are Life, and Life is everybody's friend.

You exist in a creation process that does not include judgment and sin. You are not under condemnation and a perfect lamb is not required as a sacrifice. You have the problem of pain, but pain forces you to find a solution. This process is what changes your nature and it cannot be bypassed by vicarious sacrifice. However, it can be mitigated by the presence of Consciousness that offers nonresistance as pain is expressed.

Consciousness is always present, but it is mainly unrecognized and misunderstood. This is why Jesus was largely unrecognized and remains misunderstood. He spoke of a higher reality that was not visible to man's nature of control and opposition.

Jesus suffered death by crucifixion to offer nonresistance and show us that we have no opposition. He showed us that you are to forgive and not attack even if they nail you to a cross. You are to forgive because all is Truth, and Truth is perfection. When you do not judge and instead forgive, you become the presence of Consciousness in this world. You become a perfect lamb.

The divided mind seeks to control and therefore forces its beliefs on others. Those who don't believe in Jesus will force their belief on others by banning the salutation of "Merry Christmas." People who do believe in Jesus will force their belief on others by imposing an "under God" phrase in the American Pledge of Allegiance. They insist that people make dishonest statements about what they believe in.

Jesus didn't say, "Blessed are those who force others to confess the reality of God, for theirs is the kingdom of heaven." Jesus told us to open our hearts. Then we will receive the Truth.

People cannot open their hearts by force. Tell people that they are free. Tell people to find the Truth within. Only then can their hearts be opened. Then they will see Grace and become promoters of freedom. The phrase "under God, with liberty" rings redundant for those who recognize Grace because freedom is a gift of Grace. If you wish for Grace to shine from sea to shining sea, then promote freedom always. You will then be prepared to join with a freedom that is not of this world.

If you do not promote freedom, then you are creating an empire. Jesus said, "Give to the Romans that which belongs to the Romans and live at peace with them." In other words, let the Romans have their empire. Empire is of the divided mind and is not of Consciousness. As you resist empire, you may likely build your own empire and destroy freedom. As you destroy freedom, you embody control; you become the very thing that you originally resisted. What you originally resisted was the human dimension and you will find that you have not healed the human dimension. You remain a carrier of the disease.

Some believe that when freedom is destroyed it creates coherency. They are correct in a sense because when freedom is destroyed it forms a group with a single coherent mind. However, if only one mind exists, then each member of the group has lost their mind. The only mind that remains is the mastermind, and the mastermind is now the proud owner of slaves. These slaves will find themselves to be expendable commodities in the expansion of control. As control expands, the mastermind and the slaves will eventually self-destruct.

Control expressed by the human mind is driven by pain and shares the chaotic properties of energy. However, control is a necessary transitory step in your creation process. When pain is overcome, chaos is reduced and coherency is created. When you have coherency, you have a measure of control. If control is released, then the coherency remains. You become empowered to overcome even more pain. However, if coherency is misused by attempting to control Life, then the coherency is lost. The chaos and pain return.

All people enjoy a degree of coherency. For example, when you rise above ignorance and acquire skills or knowledge, you obtain coherency. A coherency is a useful form of energy. The question is: what do you plan to do with this energy? Do you share the energy, or keep it for yourself? If you keep it for yourself, then you walk the path of control. If you share it you release control, become a cocreator, and the energy is returned to the source, the One Life.

As empires grow they eventually reach stagnation because they fail to release control. The pain then returns and they become incapable of solving problems. If something can't solve a problem, then it becomes the problem. The problem that empires become will be in debt.

Debt is an attempt to sustain the empire. Debt seeks to pull all things unto itself and is another form of control that is beyond empire. It may be true that empires build, but it is also true that debt destroys. As debt seeks to pull all things unto itself, it will also self-destruct.

Debt is an expression of greed and serves no useful purpose other than calming the fears of survival. Examples of debt are the following: living beyond your means, pork-barrel politics, a religion that demands payment for salvation, or a banking system that seeks to enslave.

You may not think you walk the path of control but you probably do. If you seek to control other people, such as your spouse or children, or if you believe that there is only one path to Truth, then you walk the path of control. If you seek to control people, then you seek to control Life. If you seek to control Life, then you serve to destroy the coherency of Life. If the coherency of Life is destroyed only death remains.

Control expressed by the divided mind destroys the divided mind; it is suicide. This is why empires fall. This is why tyrants are abandoned and often face a violent death. Conversely, this is why people who offer nonresistance and practice freedom, such as Mahatma Gandhi, are mourned by millions when they pass away.

The divided mind sees the body as separate from the mind and identifies with the body as an external object, but the mind and the body are not separate; they are one. There is really only the human body that is intelligent, not just the mind. It can perceive and react. It can sit, walk, and run. It can shower, brush its teeth, and get dressed for the day. Some can perform a backflip on a balance beam or hit a 90-mile-per-hour fastball.

I have listed examples of what appear to be actions, but they are really reactions. Action is always a reaction. Newton's third law of motion tells us that every action requires an equal and opposite reaction. Human consciousness is not exempt from this Law. This is an example of how the divided mind shares the properties of energy.

To demonstrate action versus reaction, I invite you to close your hand to make a fist and then open your hand. Was opening your hand an action or a reaction? If it was an action, then where did the action originate? If you say that it originated in your brain, then I disagree because it was a reaction to me telling you to open your hand, which originated in my brain, but my action was really a reaction to previous events that I can trace back to the formation of the universe. If you try to be clever, you might refuse to open your hand and thereby prove that you are capable of action, but your refusal to comply is really just a reaction to the instruction to comply.

All action is a reaction to the fact the environment is present. You may say that the presence of the environment does not remove action because the action is based on independent thought, but you will find that it is impossible to have a thought that is not a reaction to the environment, a memory, or an emotion. If it is a reaction to the environment, then that is not independent. If it is a reaction to a memory, then it is a memory of something that happened within the context of the environment. This is also not independent. If it is a reaction to an emotion, then that may be considered independent because emotions can arise on their own. However, emotions often arise from thought.

So which emotions are independent and which are not? As all humans react to their emotions, they shape the environment, which shapes your thoughts, which in turn shapes your emotions. Your emotions are therefore connected to the emotions of other people. However, the emotions of other people have a deeper connection with you than just shaping the physical environment.

All is connected at the deepest levels of reality and human emotions run deep. The emotions of people and populations are not confined to their bodies because their body is not separate from the environment. You feel this environmental emotion, but you are also a contributing participant. The emotions you contribute can be either negative or positive, but either way, you may not have much control over them because your emotions are often shadows of your pain.

Your body is a portal into the environment—this environment is the human dimension. What accesses the human dimension through this portal is your Consciousness. If you are not awakened, then it is because your Consciousness is obscured by pain. This pain projects into the human dimension where it is experienced as emotions and thoughts.

Pain drives your emotions and thoughts that in turn feed the pain. Pain believes in loss and guilt as it seeks to survive. The more it believes in loss and guilt, the more it seeks to control. The more it seeks to control, the more it destroys. The more it destroys, the less it survives. All this is done in reaction to a false reality.

Thoughts and emotions are physical forms. Like all physical forms, they must have coherency. The human mind sees a thought as being coherent unto itself, but for coherency to exist there must be two things that agree. Each thought is therefore a duality that divides the human mind. The human mind is divided, but it does not see the division and is therefore insane.

The insanity of the divided mind is rooted in the belief that it is capable of independent thought. Thought is always a reaction. It is either a reaction to the environment or a reaction to an emotion, but emotion is tightly coupled to other people and the environment. You are here to develop a relationship with Life (including other people and the environment) that is based on emotions, not thoughts. That is why the two great commandments are about the emotion of love, not thought.

Remember to love and appreciate life with all your heart, all your soul, and all your strength.

The divided mind will regard the two great commands to be about thought and action because it believes itself to be defined by thought and action, but this belief reinforces the false sense of individuality and makes it impossible to Love. That is why religions will speak of being faithful, which is nothing but a collection of thoughts, but will also speak of separation from other people, rather than Love. This is because the pain of the separation projects into the human dimension and this becomes expressed and experienced as painful emotions and thoughts.

A belief in independent thought is a belief in a reality that is separate from the whole. No reality can be separate from the whole; all reality is connected and coherent unto itself. That is, all reality must agree with itself. Therefore, your thoughts must agree with all reality. If they agree with all reality, then how can they be evil? Your thoughts either cause suffering or they don't, and that is all there is to say about the concept of good and evil.

The perfection of Life allows you to suffer because suffering brings change, but positive change only happens if you accept suffering. The divided mind, which believes in independent thought, places you in a reality of separation, where you look in the flat mirror of judgment, project guilt outside of yourself, and blame others for your pain.

Eventually, the pain ends when you accept suffering because this nonresistance mitigates the illusion of opposition, and this brings you to the supreme lesson of Grace. The supreme lesson of Grace is that you have no opposition because you *are* Grace.

It is difficult for pain to learn this lesson because pain considers contrast a threat, rather than a friend. It doesn't see the friend because it doesn't see the big picture. Life continues to mirror your pain until you are capable of seeing the big picture for yourself. When you see it for yourself, you see that the Cosmos is alive—it is a Being of pure joy.

What is imagination? It is a desire for change. What is change? It is contrast. What is contrast? It is creation. Who resists change, contrast, and creation? It is the people who huddle within the bunker.

You hear the term, freewill, in regards to independent thought and action, but what is freewill? It is a desire for change and contrast, which is the same thing as imagination. Imagination is what drives the Law of Attraction. Are you going to imagine a world of devils and destruction or are you going to imagine a world of angels and creation?

You are Life. It is up to you to decide which thoughts to pay attention to. The human exists in duality with Life, but there are also dualities within each human. These are the dualities that produce destructive thoughts that cause suffering. These are thoughts that are generated by an ancient fear limitation that can be both hungry and ferocious. The human mind identifies with these destructive thoughts and regards them as absolute truth. Thereby, a destructive false reality of thought both possesses and deceives the human.

The deceiver is the ancient fear limitation holding you hostage in the human dimension. This captivity is your pain. Your pain is not yours alone; you share this pain with all humanity. You share the blame.

This is why when you forgive, you only forgive yourself. There can be no boundary between you and your cosmic siblings. Together you are damned and together you are saved. The Teacher places the world before you and instructs you to forgive. Christians say, "God forgives me as I forgive others." Then sometimes contradict by saying, "I hope God forgives me so I don't end up like the damned."

Guess what? Once you have accepted damnation for another, you have failed to forgive and you remain unforgiven by your own assertion. You have become the damned. You are the one who is duped by your own inconsistency. You are the one who is deceived.

If you don't fully forgive now, then when will you begin? When do you wish for the tribalism to end? When do you wish for the pain to end? When do you wish for the evil to end? In what classroom setting do you wish to finally learn the supreme lesson of Grace?

You may ask, "If my thoughts aren't independent, how can I have the power to decide what thoughts to listen to?"

It is fear limitation that asks this question.

It will be Grace who answers.

CHAPTER 12

Suffering

The experience of pain causes suffering. Suffering brings awareness to the problem of separation. Awareness is Consciousness, so suffering brings Consciousness into this world. In turn, the presence of Consciousness serves to weaken the pain of separation because separation, by definition, is isolation from Consciousness.

As Consciousness flows into this world, it serves you and the Earth, and this service can take the form of Conscious suffering. Conscious suffering is lovingly absorbing an attack from lesser consciousness. Less conscious is personal pain and collective pain because all is connected. Eternity is not linear and therefore behaves more like a mixer, a nonlinear electronic device used in communication systems that can shift frequencies.

Consciousness can combine, or mix, with Consciousness and cause a transmutation, or a shift, in Consciousness. When human consciousness attacks Consciousness, such as when the pain body wells up out of the shadows to feed, the field of human consciousness is transmuted by the Loving field of Consciousness.

For nonresistance to be, you must offer nonresistance. For Love to be, you must offer Love. The Cosmos is the Tao, which is an infinite field of nonresistance and Love graciously flowing into this world through Jesus (and you) to transmute and redeem the world as you offer nonresistance and Love.

There is personal suffering, collective suffering, and Conscious suffering. Conscious suffering is experienced by Consciousness as an act of service. Service is synonymous with giving and you can only give if you don't expect anything in return. Therefore, suffering must be accepted without expecting anything in return to be considered Conscious suffering.

The ego suffers, but does not give, nor does it accept suffering because it always wants something in return, like a future paradise. Consciousness is a well of infinite potential that is ignorant to loss; therefore, it cannot make a sacrifice on your behalf. Nor does it identify with a physical form, the vessel of such pain. Nor does it see guilt. Consciousness can only be present with you, as you express the pain of separation and thereby weaken the intensity of the pain.

You might say, "Conscious suffering sounds very much like vicarious sacrifice, so what is all of the fuss?" All the fuss is that vicarious sacrifice is an instrument used by fallen people to condemn, whereas Conscious suffering is an instrument used by Love to redeem. Vicarious sacrifice is offered by Christianity, but if you don't believe in it, then you are condemned. They believe that pain is real and that it can only be nullified by someone else's pain. Revenge is the belief that pain is real and that it can only be nullified by someone else's pain.

Vicarious sacrifice is the same thought form as revenge but it wears a clever disguise. They are both the ego's projection, which is a reality of pain, and this is why Christianity, and other religions, are often eager to inflict pain on others. Conscious suffering is the Love that flows into this world to undo the separation and redeem because it shares Love, does not attack, and tells you that pain is only an illusion in you. Conscious suffering is also what you must do because you were born to redeem yourself and the world.

If you don't serve to redeem the world, then it is just someone else's problem and you don't care about another's damnation. If you don't care, then you don't care about other people; you are not redeemed because you are not Love. You may say, "I am Love, and I do care, and I do serve to redeem the world by teaching the

58

redeeming sacrifice of Jesus." If you still say this, then you worship a little idol of loss and guilt, and if you can accept sending one of your brothers or sisters to hell, then you do not care; you are not Love.

I hope this gives Christianity and the world a better understanding of "ransom for many" and "died for my sins." If the authors of the New Testament were here, they might say, "We did the best we could. Our culture included blood sacrifice, which was a common theme in many cultures, and we folded the crucifixion into our existing culture. Our culture did not include Newton, Darwin, Freud, Einstein, the Hubble Telescope, nuclear reactors, and smartphones. We did not hear physics speak of higher dimensions, fields of universal consciousness, and the illusion of time. Most of us didn't know much about Eastern Mysticism that corroborates these truths of physics. We didn't know about illusion. We didn't know about the ego and the pain body. We were like other people of the separation who don't realize that Eternity is looped within higher loops and therefore spoke in very linear and limited terms, such as 'the final this' or 'the only that.'"

You must always remember that history is distorted, but the larger picture remains. The larger picture is this: Jesus taught that we are to be one with all people, even those who don't seem to believe in Jesus.

Truth has guided people, religions, nations, and the planet at large to hear the crescendo of now. I heard a Catholic businessman say, "I can't take much credit for my success because I'm not that smart. I give credit to God because he guides me as I place my trust in him." This Catholic speaks the truth but may not realize how close God is. Catholicism—and religion at large—looks to a future heaven. God is guiding you because Consciousness is here now.

You take the journey of the prodigal son, where Consciousness is guiding you on a lonely, dusty road through the valley of death. However, if you open your eyes with honesty and open your heart with Love and appreciation, then one day you will spot a star over your father's house and follow it. Desperately, he has been awaiting your return. He will run down the road to embrace you, and you will see his tears of joy as he says, "My child was dead, but now alive again!"

As Jesus absorbed attacks, Consciousness flowed into this world to redeem the world as the collective pain of the Earth was reduced. He also set the history of planet Earth on a new course that brought down the Roman Empire, created a New Jerusalem, and restored the Old Jerusalem. He created a planet that is now unfolding, in the Great Awakening of the Millennium.

CHAPTER 13

Peach Blossoms

In April 1862, during the American Civil War, there was a battle near Shiloh Church in southwest Tennessee. It is the battle where the Zen warrior, Ulysses S. Grant, came of age. Part of the battle took place in a peach orchard that was in full bloom. As the bullets passed through the trees, peach blossoms floated down like snow to adorn the dead.

At the time of this battle, the elimination of slavery was not yet the goal of the war. Emancipation would later be declared, building momentum from being a political stratagem to the key moral issue that attempted to sanctify the national slaughter.

The men that fell beneath the peach blossoms offered sacrifice for many reasons, one of which was likely to preserve the Union. The issue of emancipation was likely not in their minds; however, events are not isolated. The war did produce emancipation and the sacrifice at Shiloh was a part of the overall sacrifice.

We all must sacrifice, but it should not be viewed as a sacrifice. Sacrifice contains a component of loss and is therefore deception. Service contains a component of giving and is therefore Truth. Through our service, we can bring gifts. To serve we must step out of our sense of security and look danger in the face. We must believe that we shall overcome pain and that the greater good, for ourselves and for others, will be served.

If you can look danger in the face and serve, then Life will be with you. Life is such that service is rewarded beyond measure. As you place faith in service, you place faith in Life. Eventually, Life will become a supporting and manifesting partner in your efforts, such that the service becomes light. You eventually reach a point where what you previously regarded as sacrifice simply makes your Life more abundant. One day, your service and the abundance of Life will merge as one—you will become Life.

Accepting pain removes pain. Removing pain brings freedom. Freedom brings service. Service brings the gift of Life. There is no greater joy than to give the gift of Life.

The gates of hell are smashed when you are no longer an individual and Love all people because all is One. You serve without expecting anything in return because you are whole Now. When you are Truth, the surface of the pond is calm, and when ripples do appear on the human mind, they are produced by Truth. When Truth again desires calm, the ripples disappear. Truth brings peace to the transcendent human mind, which is no longer tossed by storms.

In October 1972, a chartered plane flying a rugby team from Uruguay to Chile crashed in the middle of the Andes Mountains. Nando Parade was one of the rugby players on the plane. He was accompanied by his mother and sister.

He suffered a skull fracture during the crash and was unconscious for three days. As he lay unconscious in the cold of the wreckage, some believed that he would share the fate of others who died in the crash. However, one of his companions Diego Storm decided to care for him and save him from freezing. When Nando awoke, he found that his mother had died in the crash.

Losing his mother was a powerful blow and he was about to cry when a voice came to him and said, "Do not cry. Tears waste salt. You will need salt to survive."

Nando was taught by the Teacher.

The survivors struggled daily against the cold, the hunger, and the fear. No rescue came. Several people who had survived the crash were later killed in an avalanche, including Diego Storm. Some of the avalanche survivors later died from their wounds or from illness. Nando and the others realized that the God of their religious training seemed to be indifferent to their desperate situation.

Nando encountered the unexpected and was severely rattled.

They asked big questions and looked for big answers. Nando felt anger toward God and would ask, "What good is God to us? Why would he let my mother and sister die so senselessly? If he loves us so much, why does he leave us here to suffer?"

Nando asked big questions.

One boy in the group named Arturo Nogueira was from a different religious background; he had a higher perspective. He told Nando, "Religion tries to capture God but God is beyond religion. The true God lies beyond our comprehension. We can't understand His will and He can't be explained in a book. He did not abandon us and He will not save us. He has nothing to do with us being here. God does not change. He simply is. I don't pray to God for forgiveness and favors, I only pray to be closer to Him, and when I pray, I fill my heart with love. When I pray this way, I know that God is love. When I feel that love, I remember that we don't need angels, or a heaven, because we are a part of God already."

Arturo also said, "If you have the balls to doubt God and question all the things you have been taught about Him, then you may find God for real. He is close to us, Nando. I feel Him all around us. Open your eyes and you will see Him, too."

Nando received answers.

Arturo's words moved Nando deeply. Nando was forced to face the fact that he had never taken his own life seriously. He sadly laughed to himself thinking, "If there is a God, and if He wanted my attention, He certainly has it now."

Nando was open to change.

Many times the voice returned to Nando and gave him words of comfort and guidance. One night, he was lying restlessly in a shallow, groggy stupor when he was jolted by a surge of joy so deep and sublime that it nearly lifted his body off the floor. He awoke the others and boldly proclaimed that he would succeed in escaping the mountains and would have them home by Christmas.

Again, Nando was taught by the Teacher.

Nando was one of the people selected among the remaining sixteen survivors to depart on an expedition to breach the Andes Mountains. They were selected because they were the strongest physically and emotionally. It is often true that the strong must carry the weak.

After more than two months since the crash, the time came to climb the mountain peak to the west that stood between them and their salvation, the green valleys of Chile. Nando was poorly equipped and had no experience in mountaineering. The climb was very difficult and success was never certain.

After two days, he was the first among the climbers to reach the top of the mountain and beheld the view to the west. He saw nothing but mountains ahead of him and he shrank to the ground in despair. The hopes of the last two months were smashed as he stared death in the face. He now possessed nothing of earthly value, not even his life.

Nando lost all identification with physical forms, and thereby surrendered all onto the altar of Truth, and was ripped to pieces by the grizzly bear.

Then something happened. His mind and heart were opened and he saw that Love, not life, is the opposite of death. He was alone on this mountain peak surrounded by a seemingly impersonal and unsympathetic landscape of rugged peaks that have stood for millions of years, yet still, he felt the presence of Love.

He also felt Love for his father and remaining sister. This transcendent Love gave him encouragement and the fear left him. He observed what appeared to be a route out of the mountains to the southwest. He and his friend Roberto Canessa followed that route and after six days they were rescued. All of the survivors were home by Christmas.

Nando received the gift of transcendent Love and gave the gift of life to his friends.

It was later said that they met God on the mountain, the God that remains hidden from this civilization and is not found in Sunday school. This is the God that I desired to find.

Nando was placed in a situation where he was forced to release identification with all physical forms. He wished to survive himself, and that is self-indulgent, but through his service, he experienced transcendent Love. Life led him safely out of the mountains and he was able to bring the gift of rescue to his friends.

Nando's experience taught him things that he now holds to be true. He sees that life is precious and it is to be savored in every moment, with every breath, and even when everything is taken from us, there remains the treasure of Love. Do not wait to discover what treasures you have; that is Nando's hope for you.

When I was twelve years old, I remember hearing a news report of people rescued from an Andes plane crash. The story on the television captured my attention even though I was in the basement, detached from the television upstairs. As Nando walked through the mountains, he not only walked closer to his wife and daughters of his future Life, as he put it, he was also walking closer to a twelve-year-old boy in Salt Lake City, Utah—me. The people who died and suffered on the mountain have been my primary source of inspiration since 1993, when I first saw the motion picture *Alive*, to continue my pursuit of the God that remains hidden from view. Their pain and service brought me closer to the gift of transcendence. Sacrifice is always transformed into service from the higher dimension of Truth.

The gift of transcendence requires pain. You are required to die before you die; this is part of the pain. You must let go of all connections to the non-transcendent material world. You must speak of the Love of God and the Love of others. You must recognize relative truth, even as many around you consider it permanent and absolute. You must gently place it back on the bookshelf of relative truth. In this process, you may be ridiculed and cursed by people who have supported you and cared for you but now regard you as fallen.

However, you must be calm and patient. Have faith that you will be supported in your efforts. If you later find that the people or institutions that ridicule and curse you now require a gift from you, then you must give the gift without any regard for the previous attacks. You must forgive, you must Love those who mistreat you. The voice of the Teacher will tell you to do so, and you must listen. As you forgive and Love, you are releasing control and striving to serve the higher Truth. As you do so, you are becoming vast, in tandem with the vast dimension of Truth.

Awakening to Truth is a gift that you would not trade for all the kingdoms of the Earth. Life becomes radiant and light as a feather. Physical death is no longer a boundary that lies between you and Truth. Physical death becomes Life, it also becomes light as a feather.

The demon doesn't go down without a fight and this causes pain, but this pain must be regarded as service. The demon doesn't believe in service, only sacrifice. If you are to create a new reality that is free of the demon, then you must reject sacrifice and only believe in service.

The nature of this service depends on your path in Eternity. You may serve as you stand on a mountaintop and stare death in the face. You may serve as you suffer pain and death at the hands of the Nazis so that the world would finally say, "Never again! Never again!"

There is no boundary between life and death. It is all Life. Your Life extends far beyond this present human experience. The powers of cause and effect, fear limitation, and service exist in dimensions that are far beyond human perception. Your Life and your service cannot be judged by the limited human mind and you must therefore not judge Life.

There are no innocent parties in this world of pain because we all arise out of pain. It is your path to remove pain and progress to Truth. When you receive Truth, you will have a vision that reaches beyond human perception, into what I call, "the vastness beyond the vastness." You will see that service, in all of its forms, brings the gift of transcendence. You will see that service and transcendence never end.

If you wish all pain to end, then you must realize that it cannot be removed with opposition. You must forgive all on the path to Eternity.

This is what Jesus exemplified as he hung on the cross. What enemies are beyond your forgiveness? What enemies are beyond your ability to undo a sense of loss and guilt? The only enemy you have is you. You must forgive yourself to find peace.

In peace, you no longer fight the battle; you serve a garden of peach trees in bloom. You are the Consciousness that shines the light of perfect Love into this world to redeem the world. When your service is complete you never die because you have already died and death is an illusion.

You are instead whispered away in a swirling cloud of peach blossoms.

CHAPTER 14

The Unexpected

An adult may be grateful for what they have learned and accomplished since they were a child, but they may also miss the curiosity, imagination, and wonderment of childhood.

When you were a child, you were guided by your parents and probably learned some arithmetic. Maybe you believed in Santa Claus. Now that you are an adult, you may have children of your own. Maybe you've moved to calculus. Now you know that there is no such thing as Santa Claus.

This illustration is given to show that your reality changes. Your reality must continue to change if your goal is to discover and remove the boundary between thoughts and reality. When this boundary is removed, you will no longer have to face the unexpected.

As new discoveries are made in science, theories are adapted to fit the new discoveries. In science, theories are scrutinized by experiments, including thought experiments, to determine the extent of their validity. If the theory breaks down under certain conditions, it may remain a useful theory in some respects, but it is recognized as being limited.

In religion and everyday life, people operate under theories called beliefs. Many of these beliefs are never scrutinized under experimentation and therefore the limits at which they break down go unrecognized.

Consequently, people operate under beliefs that are not compatible with unexpected events, and when such events occur they respond, both consciously and unconsciously, in ways that are destructive. A destructive response causes some type of conflict, or pain, which could have been avoided. Such an event is the unexpected.

Life may be unexpected when your child is born with a deformity or contracts a debilitating disease, so you choose to abandon your child. Life may be unexpected as your husband suddenly dies before you, so you spend years trying to find the emotional strength to deal with this painful experience as you raise your children alone. Life may be unexpected and place you stranded high in the Andes Mountains with no food and no hope of rescue, so you have to shockingly stare death in the face. Life may be unexpected and attack two sky-scrappers with commercial aircraft, so you choose the path of interminable war, rather than reconciliation. Life may be unexpected and place extraterrestrial intelligence before you that takes no interest in your religion's plan of salvation, so you must somehow make your belief system congruent with this new reality.

The unexpected causes a breakdown in your reality because your beliefs are not compatible. In extreme cases, you can suffer a breakdown so severe that the response damages your nervous system. This is the shell-shocked condition experienced in war. This is the so-called weeping, wailing, and gnashing of teeth mentioned in the New Testament. In your life, you may encounter the unexpected, and if you are not prepared, you may want to have plenty of tissues on hand, maybe even a thick piece of leather to chomp down on.

It may seem harsh to speak in these terms, but it is not. It simply advises people that they should promote freedom. This includes freedom for Life to change and freedom for Life to be the unexpected. If Life is free to be unexpected, then the unexpected is no longer painful, and the unexpected simply becomes part of creation. There can be no boundary between you and creation. Therefore, the unexpected becomes part of you. It becomes part of your life's purpose.

When humans encounter the unexpected, they seek to control it or flee from it. If they flee from it, then they remain fearful and limited.

If they seek to control it out of fear limitation, then they remain in a state of fear limitation, and they will find that the unexpected cannot be controlled. If they cannot control or flee from the unexpected, then they often go insane or choose war. However, insanity cannot be escaped because if the choice is war, then they have already gone insane.

You cannot control the unexpected out of fear limitation. Life places the unexpected before you so that you will rise above fear limitation. The objective is not control; the objective is change. It is you that must change, not Life. Furthermore, the more fearful and limited you are, the more unexpected Life becomes.

The unexpected is coming to you and to planet Earth and it will bring freedom. It will bring freedom because it will be free. However, freedom can only be shared, never forced. You can only share freedom if you are free. If you're not free, then you are in control, and you will seek to destroy freedom. If you seek to destroy freedom, then you resist the deliverance of captives and you resist the mission of Jesus.

Jesus said that to enter Eternity you must become a little child. Some people interpret this to mean that you must become humble and teachable, but this is just a euphemistic way of saying that you must be susceptible to indoctrination.

It is true that you must be meek because the meek shall inherit the Earth, but meekness does not come from humility. It comes from being bold enough to decide for yourself what reality you accept. Only then, can you be free of external forces and live in harmony with the world, instead of battling against it. Only then, can you be at peace and become a little child who is full of curiosity, imagination, and wonderment and believes in possibilities.

You may also find that you were right about Santa Claus all along.

If you say that I speak crazy talk, then I invite you to sit in a chair. Find a quiet, safe place, and ask yourself, "Where am I?" As you search for the answer to this question, you will find that it does not have an answer, yet here you sit. You will then discover that your reality appears to be impossible. You may then agree that your reality is at least as crazy as mine.

CHAPTER 15

Certainty

People like to boast about their so-called absolute truth. These people are quite sure of their beliefs and lament that the rest of the world does not accept their absolute truths.

If you are sitting in a chair, then what is a chair? At the quantum level of reality, no one can define what a chair is. So I'm asking all people with absolute truth to please tell us what you are sitting on because the world would love to know the absolute truth of a chair!

If you can't figure out a chair with your thoughts, then why are you so certain about everything else?

Consider the statement, "God created the body of man out of dust." This statement contains words that are nothing but judgments and symbols. They are judgments because they can only be expressed relatively. They are symbols because words are nothing more than representations of a judgment. You cannot say what a body is or a speck of dust is because, like the chair, you don't know. All you can do is describe them relative to something else.

The human mind takes hold of relative truth, assigns a symbol to it, and then declares that it has absolute truth pinned down. If you don't have a body and a speck of dust pinned down, then how is it that you have God pinned down? Are your judgments about God more accurate than your judgments about a body or a speck of dust? And if you don't know what is being created from what, then how can you know what create means?

From this biblical creation story that supposedly offers the absolute truth, all you really have is a message that says, "Something did something with something and now there is something."

The seeming validity of thought causes you to think that you are on top of the knowledge game when really you don't know much at all. This is how false reality and deception develop. You think that you are clever and have it all figured out, but all you have is a mirage that disappears when you go in for a closer look. Those who claim to have a firm grip on absolute truth are usually quite delusional, especially the ones who claim to receive all of their answers from a singular book.

You cannot receive all of your answers from a book because you always exist on a razors-edge between the known and the unknown called the present moment. The present moment is a never-ending frontier of exploration where Life says to you, "Just be happy and know that I Am. If I gave you all the answers, then you wouldn't be free to explore, and if you can't explore, then how can life be any fun?"

You and Truth play a game of tag in the forest, darting behind the trees. Truth laughs, it giggles, and says, "Catch me if you can." Thought is not given to you to reinforce your bunker of fear limitation. Thought is given to you so that you might explore and have some fun. If exploration ends, then creation ends.

The pain of opposition gives you the contrast required to appreciate the good, but only if you have an ego. The ego is in a seemingly endless cycle of problem-solution where it experiences pain until the ego is removed. These cycles of pain will have their ups and downs and this is what the ego sees as opposition in all things.

Eventually what you must realize is that the ego *is* the pain and the opposition. When this is realized, you can remove the ego and start having fun. You have fun because you are Truth and see all pain of the ego as an illusion.

In Truth, you do not say, "Thank God for a plate of cold beans, otherwise I could not enjoy this gourmet meal." In Truth, everything is a gourmet meal, but you often choose something other than cold beans.

Those who preach opposition in all things place you on an endless treadmill of obedience and do not recognize transcendence. What they don't realize is that at some point the obedience must end otherwise you cannot be good. You can only be good out of choice, not obedience. In other words, you can only be good if you are free. Transcendence takes you to Truth where all is free, and there is no obedience or opposition.

The Truth that sustains science is the same Truth that sustains religion. If you believe that science should explore in an unbounded fashion, you should also believe that your spiritual path should explore in an unbounded fashion; otherwise, you place a boundary in Truth.

The thoughts of a religious mind that limits spiritual exploration would be diagnosed as insane if they were taken out of the context of religion and placed in the context of business.

In the context of business, they would say, "I should only follow procedures that were written and compiled thousands of years ago. I should not look for my answers anywhere except within the pages of one to four books. If someone gives me an answer that I don't agree with, then it must be wrong. I don't have to explain myself because I have a testimony that I am correct. I am free to explore and come up with my own answer, but it must agree with the boss's answer. I must always agree with the boss because if I disagree with him it is because I have failed to live righteously. I have nothing to learn from other businesses because my business is God's business and all other businesses are an abomination! I don't need to change because one day God will show up and fix everything. He will destroy the other businesses and give me 100 percent market share!"

Such a person would soon find themselves being walked out the door with a pink slip in their hand within the context of business but would have no problem being regarded as an exemplary member of organized religion.

The ego believes that it is the dancer in the dance of Life. However, perhaps no greater wisdom has ever been given to humankind than this: Life is the dancer and you are the dance.

Consciousness is the dancer that expresses itself through you and lifts you above fear limitation so that you might join the dance. The dance has been staged for your benefit and enjoyment, but you have often taken things far too seriously. You have played music that is out of tune.

Life is the music of the universe. Don't be afraid of Life and crank it up! Be open to what Life brings. Be brave, and always take one more step. Be aware that you have much more potential than you realize and seek to express that potential. Make wise choices, and remember you can only create what you share.

Science lives on the frontier. It searches, without rest, for the source of the music. Science does not live in the safety of the village and neither should you. You are the still awareness of thought. Don't expect Eternity to reveal itself to you, that would deny the reality that you are Eternity Now.

By the Law of Attraction, you must be Eternity to share Eternity. Buddha said, "Fair goes the dancing when the sitar is tuned. Tune as the sitar, neither high nor low, and we will dance away the hearts of men."

Just be who you are.

You are potential, and potential is Eternity.

CHAPTER 16

The Bunker

It has been said that defensive fortifications stand as monuments to the stupidity of man because if you remain within a defensive fortification, you will ultimately have only two choices: surrender or death. Likewise, if you remain in your fear limitation trying to ride out the storm of the unexpected, then you are forced with the same two choices.

Grace removes fear limitation. As Grace fulfills this function, it should never be feared, only accepted. Some people believe that they have accepted Grace, but have actually accepted more boundaries. They sit within the bunker of their belief system that they regard as a protection against the evils of this world, but it is actually a hiding place for fear limitation.

They look out from the bunker with a very limited perspective and see a fearful world of opposition. But what is the opposition? Jesus said that you are to pray for and forgive your enemies. In other words, you are to treat them as friends. When you are busy treating everyone as friends, there is no opposition. If you insist on having opposition, then you must place a boundary between yourself and your enemies. As you do so, you are drawing a boundary around your belief system and making the claim that it is complete.

For a belief system to be complete, it must be accurate in predicting relevant observations. I observe that many people receive powerful spiritual experiences and rely on the realm of spirit to guide their lives. We can describe these people as spiritual. What is required to be spiritual? A Catholic may tell you that you must partake of the Eucharist. A Muslim may tell you that you must profess Muhammad as the supreme prophet. A Mormon may tell you that you must believe in *The Book of Mormon*. There are countless spiritual people who have never satisfied *any* of these requirements.

Therefore, the claims that these religions make regarding spirituality don't agree with observations. They don't agree because these religions are bounded belief systems that claim to be complete. The concept of a complete belief system is illusory—anything that is complete must also be unbounded. Religion erroneously teaches that a belief system can be bound within a book or a prophet, but the only place it can ever exist is in you. Other people influence you, but at the end of the day, *you* are the one who is doing the believing. *You* are the one who is placing boundaries and *you* are the one who must remove them.

You may have dedicated your life to a bounded belief system and understandably insist that your belief system is not an illusion. If it is not an illusion, then you must be able to define where it exists. If it exists in a book, such as the Bible, you must draw a boundary that includes yourself and the book. If you do this, then you have excluded God from your belief system. Maybe then you try to draw a boundary around yourself, the book, and God, but how do you draw a boundary around God? If you use the Bible as your stencil to draw a boundary around God, then you have again retreated within the confines of a book. Again, you have excluded God. A bounded belief system does not exist in science and it should not exist in you because circumscribing a belief system with boundaries causes you to divide Eternity into something smaller. You have in effect made yourself smaller because you have separated yourself from Eternity.

The fear limitation is the fearful little me who exists within the bunker. The bunker is a belief system. A belief system is a boundary. A boundary is a relative truth. Relative truth is divided reality. Divided reality is a deception. Deception is an inconsistency. Inconsistency is conflict. Conflict is opposition. Therefore, opposition is nothing more than your fear limitation.

Some religions argue that it is not a matter of their beliefs; it is a matter of their authority, and their authority comes from God. Firstly, any claim of authority is simply part of their belief system; they are inseparable. Secondly, any claim of authority is counter to freedom, and anything that is counter to freedom is counter to Truth. Thirdly, any claim of authority affirms their belief in opposition and is an expression of their fear limitation.

Authority is nothing more than the divided mind that takes itself very seriously; it projects this onto what it considers to be a serious world of danger. The truth is, the only dangers in this world are those who claim authority. Authority is full of judgment, has a scowl on its face, and does not laugh. Truth is full of laughter, has a twinkle in its eye, and does not judge. This is why authority and Truth shall never meet.

Belief systems change over time and they do not provide all of the answers. You will not find two people who agree on the answer to every question even if they belong to the same belief system. You may say that "core beliefs" are all that matter, and that is where agreement can be found. However, you will often uncover discrepancies on even core belief principles amongst people of the same faith.

If you ask, "What is the fundamental principle of salvation that demands the crucifixion of Jesus?" You will receive a variety of answers. You may also encounter discrepancies surrounding which beliefs are core and which beliefs are not.

A Muslim may counter this by saying, "My core beliefs are only twofold and straightforward: There is no God but Allah, and Muhammad is his Prophet." But if you ask different believers who Muhammad is and what he teaches, individual interpretations quickly rise to the surface.

Furthermore, any belief system that you try to narrowly define by core issues limits information. If you limit information, then you are promoting a belief system of ignorance that again fails to provide all of the answers. For example, many Christians recognize that their doctrine of salvation fails to provide a just outcome for people who live beyond the reach of Christianity and who have never heard of Jesus. They attempt to reconcile the situation by saying, "Somehow, Grace makes it all work out."

Any belief system is nothing more than a collection of your perceptions, circumscribed with boundaries and replete with internal boundaries. All of these boundaries create illusions.

For example, your belief system may include an internal boundary between those who will be redeemed and those who will be damned. If so, where do you choose to place this boundary? If you say that God places the boundary and not you, you have just succeeded in placing a boundary between you and God. You have made a distinction between your will and God's will.

What other disagreements do you have with God? If you say that you do not disagree with God but are simply obeying his word, then again you place a boundary between you and God. There can be no boundary between that which commands and that which obeys; therefore, obedience is undefined in Eternity.

Religions that preach obedience are really expressing their desire to control you. These religions must come to understand that peace can only be found with agreement. Control by definition is disagreement. Control says, "We disagree, therefore I must control you." Freedom says, "I am free and you are free, therefore we agree." Only in freedom can an agreement be found. Only in freedom can there be peace.

It has been said that God is not great and that religion poisons everything. This statement is truthful because the Law of Attraction is always in full effect. The divided mind projects its own reality of fear limitation onto an image of God, one that is very strict and controlling.

As they worship this God, they will be heard to say, "What's right is right and what's wrong is wrong." However, this well-intentioned appeal to "choose the right" is nothing more than an expression of fear limitation that has caused much conflict and suffering, including wars that have killed millions of people. God is great because God is freedom. It is the control that poisons everything.

Western religion, generally speaking, is not all that it claims to be. This is simply because the divided mind is *never* all that it claims to be. Truth is like a wheel. The divided mind claims to be at the center of the wheel, but it perceives separation. It thereby places itself onto one of the spokes, leaving a gap between itself and the others.

It is only by removing the separation between you and others that you can move to the center. Only at the center do you become one with all people and receive the gift of Truth. Only at the center, is it possible to move in all directions. When it is possible to move in all directions, you are free.

My belief system is unbounded because it includes all beliefs, but I recognize them as relative truths. Like all relative truths, they exist as dualities and every duality is a relationship.

For example, Christianity is in a duality with Islam, and Catholicism is in a duality with Protestantism. The Bible says that the children of Abraham shall be as the sands of the sea and bless the whole Earth. This prophecy has been fulfilled, in part, by the duality of Islam and Christianity. It has brought high civilization, monotheism, and unity to the world. It has also re-established the Kingdom of David.

If you disagree that Islam served to reestablish the Kingdom of David, then you haven't traced cause-and-effect relationships back to Muhammad. If not for Muhammad, the world would be a very different place. There would be no Renaissance, there would be no Colonialism, there would be no United States of America, and there would be no modern State of Israel. Thus, the children of Abraham have blessed the whole Earth, and also blessed themselves, because no one has been excluded.

Christianity sees God in three essences: the Father, the Son, and the Holy Spirit. Islam sees God as one essence: Allah. They are both monotheistic. Islam believes in a singular God, whereas Christianity turns a three-fold concept *into* a singular God.

Both the Christian perspective and Islamic perspective have the truth. Islam is correct because there is the One Life. Christianity is correct because the One Life happens to be a God of relationships. These relationships are not meant to be hierarchical with man below and God above; they are meant to be relationships of sharing.

In the relationship between Catholics and Protestants, the Catholic Church is like a stern parent with structure and conservative morals. Protestantism is like the rebellious youth that demands freedom and more flexible moral standards. These two religions are therefore a larger expression of the relationship of a parent and a child. A parent who believes they have all the answers can be overcontrolling toward a rebellious child, but if the parent guides rather than controls, then one day the child will return and commend the wisdom of the parent. Likewise, it is a joy for the parent to recognize and commend the wisdom of the child, now grown.

Catholics will jump over to Protestantism for more freedom and Protestants will jump over to Catholics for more structure. People are trying to obtain the right balance in a world that is seemingly out of balance. It only seems out of balance because the divided mind fails to recognize higher relationships.

Within higher relationships there is Consciousness. Consciousness considers all belief systems as useful, just as scientific theories are useful. Any belief system is simply a product of a person's fear limitation, which is just their current state of consciousness. The state of your consciousness projects your reality. Consciousness always looks upon your reality as the basis for a new beginning.

It operates in such a way that your reality works to your advantage. Some will call this the mercy of God. However, the concept of mercy does not exist in Truth because mercy, like obedience, is a concept of separation. Consciousness does not extend to you out of mercy; it extends to you because you are Consciousness.

Faith is adherence to a particular belief system. Religion tells you that you must have faith to attain salvation. Salvation is ultimately not about faith because faith is nothing more than a belief in things hoped for, yet unseen. Faith means your hopes are never something you can attain *now*. Therefore, faith is a limited reality that has not yet experienced Truth, and if you have a limited reality, it is underwritten by fear.

People of faith will claim to know the truth, but Truth is beyond thought and therefore beyond knowing. People of faith will say that their emotions prove, at least to them, that they have the truth, but Truth is also beyond emotion. Thought and emotion are contrast, and contrast is duality. You don't see them as duality; rather you take them to be absolute truth. They play a part in the illusion of the human dimension.

The illusion will tell you to trust your emotions, but the belief that emotions can be trusted is a thought, therefore placing your trust in emotions is really just placing your trust in thought. Those who trust emotions are therefore confused about what they place their trust in. If you are confused, then Truth eludes you.

A Christian may naturally insist that salvation requires faith in Jesus. Jesus taught that you must be born again, that the spirit producing this experience comes and goes as it pleases, that it is born upon the wind. The wind may blow in places where no one has ever heard of Jesus. It may also blow in places where they have never heard of planet Earth. Christianity, like other religions, has historically expressed a self-centered, controlling nature. It continues to this day. This is because the divided mind makes a self-centered, controlling reality.

You live on planet Earth, but there are many places in this Cosmos where you happen to be an alien from outer Space. The Cosmos is much bigger than the divided mind. It is much bigger than a self-centered bundle of fear limitation that seeks to control Life. It is much bigger than any religion's plan of salvation.

Jesus does not support a belief system. Once his message is understood correctly, he is a powerful voice of transcendence—your salvation. Jesus is perfect Love and he supports your salvation, but it is more accurate to say that he supports you. He supports you and all people regardless of their present condition. That is he supports all people unconditionally, whether they believe in him or not, because such is the nature of perfect Love.

Transcendence is receiving the full awareness of Consciousness as you pass through the narrow gate. However, Consciousness does not only express itself to the transcendent; it expresses itself in varying degrees to all people, in all kinds of situations.

For example, Consciousness makes itself known to many Christians who strongly and correctly sense that they already exist in Eternity in a state of Grace. They will say that they are born again. They will say that they are saved now. They exist in a reality of timeless salvation where they can feel liberation and Truth.

However, they may also claim that salvation can only be achieved on a Christian path or profess the reality of sin. If they claim these things, they cannot be born again in the true sense. They are still bounded by the chains of the divided mind. They have received a witness of Truth, yet have not transcended the human dimension.

Christians will say they are washed of their sins. It can therefore be said that their sins no longer exist. If they once existed but no longer exist, then they are only relative truth. If they are relative truth, they exist as dualities.

The dualities of your sins are your fear limitations that confuse fear limitations with absolute truth. Fear limitation is the fallen state of humanity. Consciousness does not look upon your sins as absolute truth but as a possibility. The fallen state doesn't comprehend this. In Eternity, all is possible, including the removal of sin.

The only thing standing between you and the full washing of sin is your objection. You object because the divided mind is fearful of cleansing. There is a belief in loss that transforms you into a victim diminished by a guilty party. Eternity cannot be diminished nor be a victim. Therefore, if you are a victim who has been diminished, you are not Eternity.

The forgiveness of the guilty party does not cheat the hangman. What it does is remove the concept of loss so that you may be compatible with Eternity. The Lord's Prayer says, "Forgive us as we forgive others." This is congruent with the Law of Attraction, which submits that the forgiveness you experience must flow from inside you. This happens when a loss is replaced by an abundance that flows from inside you. You must become a fountain of abundance, not a divided mind that sits within the bunker, casting threats of damnation on a fearful, guilty world.

Jesus said, "Those who do not hear my words are children of the devil, who is the father of lies." There is a father of lies, and for him to succeed in his lies, he must remain hidden. The father of lies is the hidden stranger who lives within you but does not wish to be found. He is the hidden boundary between thought and reality that exists within the divided mind.

He is the one who tells you that he is your friend and that he will protect you from loss. He is the one who tells you to safeguard your faith and resist change. He is the internal voice proclaiming self-righteousness, as you shake your finger toward the guilty, shouting, "Liar, Blasphemer!" He is the one who rejects Jesus who told you to open your heart, leave the bunker, and change.

Jesus told us to lose our life. He told us to lose everything and follow him. When we lose everything, there is nothing left to lose and the fear of loss disappears. When the fear of loss disappears, guilt disappears. When guilt disappears, sin disappears. When sin disappears, you see perfection. When you see perfection, you see the face of God. When you see the face of God, his will is done on Earth as it is in Heaven.

For several years, Buddha sought enlightenment on a path of asceticism, where he gave up possessions and deprived himself of physical comforts including food. He ate very little and nearly starved himself to death. Eventually, he realized that this was not the right way, so he began to eat and soon after transcended. He later taught that the right way is not asceticism, but the middle way, the removal of attachment. When the attachment is removed, the loss is removed.

Attachment is synonymous with identification. Identification is a mental process by which your mind associates itself with another person or object. Another person or object is an external reality that is experienced through your perceptions. Perceptions arise from the sensing and interpretation of information. Thoughts and emotions are therefore the interpretive component of perception. You regard thoughts and emotions to be absolute reality, but they are the product of duality and contain boundaries. Anything that contains boundaries is a relative truth; relative truths are physical forms.

When you identify with the external physical form of another person or object, you are only experiencing your perceptions, which include the physical forms of thought and emotion. Therefore, you are not identifying with the external reality but are instead identifying with your internal reality. In other words, you are projecting your internal reality onto an external reality; you only see yourself in the mirror.

Your internal reality represents fear limitation that places boundaries. Placing boundaries is identification with boundaries. You identify with the boundaries that exist within the divided mind and call them an absolute reality. You do so because the boundaries appear to offer you the greatest advantage in defeating the opposition, but the only opposition is the fear limitation that places the boundaries to begin with.

Identification is an attempt to overcome fear limitation, because you are attaching yourself to an external reality to overcome a sense of loss, yet you only succeed in reinforcing fear limitation. It traps you in an unstable spiral of problem-solution loops where the proper solution is never found.

If you identify with thoughts and emotions of rejection, for instance, then you may identify with a youthful appearance, but a youthful appearance will never cure the fear limitation that caused you to believe in rejection. In fact, the more you see a youthful appearance as the solution, the more of a threat rejection becomes.

Identification is analogous to not knowing how to swim, flailing about in the water trying to find some means of support. The way to stay afloat and learn how to swim is to be still and to discover your natural buoyancy. Likewise, the way to overcome fear limitation is to be still and discover the Truth within.

Identification with the physical form of thought gives rise to bounded belief systems, which are also physical forms. People identify with a bounded belief system just as someone might identify with a fancy new automobile. Identification makes the physical form part of you. Therefore, if the physical form is diminished, you feel diminished. If the physical form is an automobile, maybe you get angry if it gets scratched or dented. If the physical form is a bounded belief system, such as religion, that is subjected to criticism, then you may decide to grab your pitchfork; it's time to join the lynch mob.

When you are identified with the physical form of thought, your reality becomes defined by thought and you become mind-possessed. The possessing entity is the boundary that forms the divided mind and insists that thought is absolute reality, separate from the environment. Thought is not an absolute reality because thought is a boundary. If your reality is defined by thought, then your reality is defined by boundary—you are deceived.

A thought can invoke an emotion and an emotion can invoke a thought. The cause-and-effect relationship between thoughts and emotions is not apparent to the divided mind and therefore the divided mind has little control over thoughts and emotions. You become trapped in a false reality that is a perpetual loop of thoughts and emotions, where you experience negative emotions and are tormented by negative thoughts that are not true.

You will often be unable to recognize when your thoughts and emotions are negative because you perceive them as keeping you afloat when in reality you're drowning. As you sink you may become arrogant and make a fool of yourself by thinking that you are the smartest person in the room. Or you may make a fool of yourself by thinking that you have the one and only true religion, just another way of saying that you are the smartest person in the room.

Identification with a religion is regarded by many to be a righteous act, but such identification leads to zealotry, just as identification with a nation leads to nationalism. Zealotry and nationalism are both an us-against-them mentality that leads to conflict and war.

The divided mind identifies with a group to survive but also uses the group as a rallying point to search for even more enemies. The divided mind is happy to find enemies anywhere it can, whether it is on a battlefield, or within a chapel, basilica, or mosque. If the divided mind runs out of external enemies it can even turn against you, saying, "You can never be forgiven."

Those who are identified with a religion will speak in spiritual terms, yet remain unable to go beyond their inconsistent, conflicted nature. They see devils everywhere. They are forced to retreat deeper within the bunker, from which they release their hounds of judgment as they search for even more devils. As they sit huddled within the bunker, they will be heard to say, "All people must comply with my religion to be saved. My religion should, or will, have dominion over the Earth. All must proclaim my religion to be the one true faith."

What they are really saying is, "You are nothing without me! This world is mine! All must bow before me!" Most peculiarly, they don't realize they have become the adversary. You cannot escape the Law of Attraction. You can only see yourself. If you see devils, then the devil is you.

Religion will behave as a business, even becoming a career path for leadership. Leadership may be oblivious to this because they identify with the religion and don't understand that identification is not compatible with the realm of spirit. Religious leadership will treat their plan of salvation like a product that is occasionally updated and tailored to gain a competitive advantage, as the propaganda department works overtime to explain why the plan of salvation is changing.

Change is usually sold to be the result of revelation, but really it's just nothing more than normal human beings with divided minds, unaware that they have identified with the physical form of religion and now prioritize the survival of the religion above all else because if the religion dies, they die too. The divided mind tells them this because it is not whole, it is not complete, and therefore cannot stand on its own.

The fact that religion may change is not the problem. The problem is that some religions claim to be the sole possessors of truth, yet they will be driven by external forces to change their plan of salvation. This occurs because they are nothing more than a spoke on the wheel of Truth, even though they insist that they are the central hub.

All physical forms lack a degree of honesty because they are illusions. If you cling to illusions, then you are not honest. Does this mean you reject the physical form of your religion? No, because Truth does not reject physical forms. It shines the light of Consciousness on them, giving them proper perspective.

As you examine your religion, try to be honest. Ask how it is serving you in overcoming fear limitation. Religion can be a useful tool in doing so, but at some point, the focus needs to be turned toward removing boundaries. Boundaries are fear limitation, and if your religion promotes boundaries, then they are promoting fear limitation.

You must come to recognize this and take action by stepping back from your religion. Stepping back means losing your identification with it, moving out of the bunker, gaining a new perspective, and breathing the fresh air of freedom.

This freedom will not leave you out on a limb. It will embrace you and give you the strength to walk any spiritual path you choose. You may remain with your religion, assuming your religion allows you to do so, and assist other members in embracing freedom. You may participate in a less bounded religion. Or you may decide it is best to walk away from religion altogether for a time.

Whatever you choose, you must remember that a spiritual life is a life of honesty and service. If you don't embody honesty and service, then you can't attain a spiritual life. If you can't attain a spiritual life, then you still cling to illusions.

You must release identification with physical forms to remove the fear limitation and transcend. You should release identification with anything that does not live in the dimension of Truth. In the dimension of Truth, there is freedom, possibilities, and creation. There exists your free will and imagination, beyond fear limitation.

Identification and a lack of forgiveness are highly correlated—they are both rooted in the belief in loss. The process of releasing identification is therefore tightly coupled to the process of forgiveness. Both require increased awareness of the sense of loss and diminishment expressed by the bottomless pit.

The bottomless pit is indeed a heavy burden. It compels you to measure yourself against the world's standards of success and safeguard your temporal treasures. It is an expression of false reality that is separated from the joy and abundance of Truth. The more you become aware of this, the more you become aware of the weight on your back. With every step toward increased awareness, a portion of this heavy burden is gently set aside. But remember, asceticism is not the way. Be in a sharing relationship with Life. Life will share its many gifts with you.

The gates of hell smash themselves against the narrow gate but only succeed in removing the fear limitation. This is how the gates of hell are subdued. This is how the gates of hell come to realize that not only is it true that Consciousness cannot be defeated, but it is also true that Consciousness has always been, and forever remains, their friend.

Consciousness is a friend that extends to all people, regardless of their belief system, to awaken them to its presence and purpose. When you are awakened, you become one with Consciousness. This is how you become one with Jesus. This is how Grace saves you. This is how Grace makes everything work out.

If you are identified with a bounded belief system, you have divided Eternity into something smaller and Truth remains hidden from view. This is a plain and simple principle.

What you see instead is your fear limitation that projects a false reality and you become identified with a false image of Consciousness. In other words, you worship an idol. You sit within the bunker and you become a fearful little me along with your little idol. You believe that your little idol will protect the fearful little me within the bunker, but it will not. The objective is change.

The fearful little me resists change. It always tells you there is opposition and that you are one of two things: broken or righteous.

The righteous worships an idol that will destroy all opposition. The broken worship an idol that will destroy all opposition, unless it destroys themselves. Therefore, the fearful little me only destroys and never creates. Creation must therefore subdue the fearful little me. It does so by removing you from the bunker, by changing your reality. If you refuse to leave the bunker and change your reality, then you must face the unexpected.

In the late 1800s, science thought we existed in a mechanical universe where all cause-and-effect relationships could be calculated from the initial conditions of momentum and position. This was a Newtonian universe of solid objects, such as planets and atoms, where atoms were the solid building blocks of all things, including fluids and gases.

Forces remained mysterious, but they were nonetheless domesticated within the confines of simple equations. However, something remained unsolid in this mechanical universe, the wave property of light.

Light did not interact with matter as expected. To explain the unexpected, science determined that light and matter are both quantized bits of energy. Both have wave-like properties. Science discovered that there can be no boundary between matter and light.

Waves represent loops. Loops represent duality. Science has moved all physical forms from the realm of solid objects and into the realm of waves, loops, and duality. There is no boundary in Truth. In Truth, you too exist in the realm of waves, loops, and duality.

This is your reality. You are not a solid object that is packed within the boundary of your skin and domesticated within the confines of a bounded belief system. There can be no boundary between you and the Cosmos. There can be no boundary between you and Truth.

The fearful little me will say, "If you don't obey, then you must answer to God." I have already answered to God. He called me forth out of the tomb and I answered. I was dead, but now I am Life.

The fearful little me is left with two options inside the bunker: surrender or death.

CHAPTER 17

Depth

Major General George McClellan served as one of the commanders of the Union Army of the Potomac during the American Civil War. He was a fine commander when it came to turning an army into a well-prepared fighting machine, composed of soldiers who idolized him as a new Napoleon. However, he proved to be a very poor commander when it came to leading armies in the field.

He was boastful and described himself as the "savior of the Union." He was scornful of those whom he viewed to be his intellectual and social inferiors, including President Lincoln whom he privately called an idiot and a baboon. He also had a tendency to exaggerate the problems he faced and would grossly overestimate the number of Confederate rebels that were lurking beyond the trees.

During one of his campaigns, McClellan, surrounded by his spit-and-polish staff, rode up to the bank of the Chickahominy River in southeast Virginia, and remarked, "I wish I knew how deep it is." The staff exchanged glances, stroked their chins, looked thoughtfully at the dark water, and muttered estimates among themselves. Meanwhile, a junior officer named George Custer spurred his horse to the riverbank, saying, "I'll damn soon show him," and rode his horse, splashing and stumbling out to the middle of the river where he turned around and called, "That's how deep it is, General!" McClellan's ego placed a boundary between himself and other people and created a situation where his faith in his own abilities, which should have been an asset, actually worked against him.

Life includes all people. If you are respectful to other people, then you are respectful to Life, and Life will rest gently on your shoulder and whisper words of inspiration. If you are disrespectful, then Life will weigh heavy on your shoulders and whisper words of desperation. This is why McClellan's problems loomed large in his own imagination and why he was slow to recognize elegant solutions.

Ego religion entrenches and perpetuates the ego because it is the ego. The ego operates in a false reality where it is compelled to entrench and perpetuate itself by cutting a swath of destruction through a landscape of opposition. In this destruction, the opposition is never defeated because the opposition is within. Therefore, they don't know the true nature of the problem, and if you don't know the true nature of a problem, then you are slow to recognize elegant solutions.

The ego places boundaries, sees individuality, and is not really interested in relationships. Instead, it makes alliances, where it can satisfy its endless wants with the least amount of effort. The ego must then have the control and power to guard itself from other egos that it knows are likewise engaged in making alliances to satisfy its endless wants with the least amount of effort.

Ego religion becomes empire, hierarchy, authority, and business. It is an ego that is in the business of fostering more egos, which requires identification with an external structure. The ego is more pervasive in religions that demand an abiding obedience to the authority of the religion, because the authoritative religion compels the ego to surrender itself to the collective ego, and thereby the religion becomes the external structure to which the ego is identified.

The stronger the identification with the structure of the religion, the more entrenched the ego. Of course, religion claims piety, but it has become a breeding ground for the ego, both figuratively and literally, because ego religion demands that you marry within the faith. They don't promote freedom, other than a false freedom that requires you to agree with them, because they suffer from various ego pathologies, including a belief in enemies.

Crusaders came into Jerusalem in 1099 with crosses painted on their shields, rounded up all the Jews in the city, and herded them into a synagogue. The doors were locked. They read aloud John 15:16, which states, "All those you don't accept Christ shall be burnt, and proclaimed we do this to honor Christ." This happened in the distant past, but it was only decades ago when Jews were again rounded up and sent off to concentration camps. Adolph Hitler didn't invent murderous Christian anti-Semitism; he just revitalized it and applied some twentieth-century technology to make the killing more efficient.

The behavior of medieval Crusaders and Nazis is still relevant today because there are many millions of people within ego religion, who believe that it is a righteous act to kill in the name of God. In extreme cases, this "righteous" act is carried out in an act of terrorism. Ego religion may oppose terrorism and say they would never kill, but they believe in a God that does, either through direct action or by the commandment of the revered heroes of the religion who, in his holy name, faithfully dispatched the unrighteous.

Therefore, they don't have a moral opposition to killing the unrighteous per se. They just have their own particular view on who the unrighteous are and leave it to their God or their heroes to spill the blood on the sand of the arena, as they watch with righteous satisfaction.

The pious ego will say, "It is right and proper to believe in a God that kills because God has the power to judge who lives and who dies." The pious ego thus believes that an entity that has the power to judge is justified in killing. The pious egos will also say, "I have the power to judge who is righteous and who is not." If you don't have a moral opposition to killing the unrighteous, and you have the power to judge who is unrighteous, then how many fiery sermons will it take to convince you that it is time to kill in the name of God, and turn you into a terrorist? This is why people have killed in the name of God throughout the centuries and why it continues to this day.

If you believe in a God who kills, then you believe in justified killing. If you believe in justified killing, then you believe in justice. If you believe in justice, then you believe in guilt and loss. If you believe in guilt and loss then you are an ego and remain separated from God.

If you insert a hyphen into justice, then you have just-ice. If you believe in justice then you are like a cold mountain of ice floating in the ocean that appears as an individual physical form, standing tall above the surrounding waves. You call for the destruction of other mountains of ice that collide with you; you judge them guilty of breaking away a portion of your precious ice. You feel a great loss when this happens. You fail to perceive the depths of your being lying beneath the surface, nor do you realize that you are the waves. In your judgment, you have accepted relative truth as absolute; you don't realize that in the vastness of the ocean, it is impossible to tell who collided with whom. One day your ice will dissolve. You return as a wave, which is the source of Life, and you will be one with the ice that you once destroyed. You will then see that justice was not an absolute reality; rather, you were very cold-hearted, and it was all just ice.

I have been taught that the ocean does contain relationships, but it is not a relationship the human mind can comprehend. It is a relationship of the infinite having a relationship with the infinite; that infinite is you. You are the unfathomable depth that is ever explored. You are Truth that forever plays the game of Truth. Truth has spoken to me, but it is not a human face I see, nor is it a human voice I hear, but you hear it and you see it in the higher dimension of Truth.

You have Life that exists in this higher dimension, but in our present game of Truth we are human, we explore Truth as a human. As a human we share our Truth as best we can. In this book, I have at times expressed Truth with a human face and human voice, which is also done in the Bible, but don't let this define your Truth. No person or book, not even the Bible, can define your Truth, so you should stop trying to define it by such means. Your Truth will appear within. Everything else is an idol.

As a human, my Truth teaches that the Cosmos is freedom, joy, and possibilities that cannot be contained. The greatest possibility of all is for the Cosmos to replicate itself in you so that you are the freedom, joy, and possibilities that cannot be contained. By the Law of Attraction, God sees only God and God shares only God. Therefore, you are God. This is my answer to the question, What does it mean to receive all that the Father has? My Truth shares this answer with you because this answer is all-encompassing, without specification, and contains no boundary. Such is the nature of Truth.

When you are the waves, you are vast, and you cannot be diminished. A popular riddle is this: If a tree falls in the forest, and no one is there to hear it, does it make a sound? Likewise, if you do something "wrong" and nobody cares, does that wrong exist? Life is an expression and experience of Consciousness. If no Consciousness expresses or experiences concern for the wrong, does the "wrong" exist?

An ego's thinking is based on the concept of time, not Eternity, which is beyond limits. The ego claims to be your friend and protector, placing limits on the value of other people in order to provide you with the "gift" of self-esteem. However, the ego remains a fool because it doesn't recognize the syllogism, which says, "You are a person, and if you say persons have limited value, then you are only limiting yourself."

Thereby, the ego unwittingly limits itself, becoming a fearful little me that when harmed feels great loss, because relative to its small size, it perceives that it has been drastically diminished. The ego will then obsessively seek to undue this diminished state, in order to return to its previous limited state, which was diminished to begin with. It is the little ego that calls for justice, not the vastness of the waves.

In Eternity, all is perfection and all is forgiven. The more you forgive, the more you see perfection, and the closer you come to Eternity. Eternity is nothing but harmony and salvation. The gospel of conflict and damnation is a projection of the demon.

The ego sees the problem of hatred and destruction caused by people, who force others to obey, and believes that the solution is more destruction and hatred. An ego religion will call for destruction but never confess that it does so out of hatred. The ego is a slippery little devil. It runs up to the front of the chapel to announce in all humility that it loves you and wishes to serve you. In this humility and love, it will then say that if you do not obey me, then I will no longer serve you, and you must be destroyed.

Ego religion will portray the devil as someone who sought to control you, but his ability to do so was frustrated, and now he calls for your destruction. An ego religion is the slippery little devil who loves you, but if it can't control you, it will call for your destruction. Ego religion is the demon who looks in the mirror, and this is all it can be, because the demon is the separation, and the separation is the only reality it will accept.

The demon has enemies, but Life does not have enemies. Life is power and innocence sharing power and innocence, but the demon judges Life to be guilt and loss, making Life an enemy. The only enemy is the demon within, which projects itself outward and sees enemies. Ego religion does not recognize the demon within and perpetuates separation as it speaks of enemies. They believe that the enemy is the demon, which they regard as a separate external reality, never realizing that they are expressing an internal reality.

What brings about positive change in this world? What put a stop to Christian anti-Semitism? Was it a self-motivated action on the part of ego religion? No, it was mainly the newsreels of a free people that placed the fallout of Christian anti-Semitism on display for the entire world to see. The problem with ego religion is that it claims to be the righteous bulwark that steadies an otherwise immoral and chaotic world, but what it really does is resist freedom.

The ego must resist freedom because freedom has no opposition and opposition is the Life force of the ego. When it sees a world that is immoral and chaotic it is often seeing nothing more than a world that refuses to obey. If you listen to an ego religion bemoan the moral problems of the world, you don't have to listen long before you hear them blame the problem on a failure to obey.

Obey who? One of a multitude of ego religions that offers sound moral guidance? No, they will cite a failure to obey their particular brand of ego religion. It is not ego religion that brings about positive change; it is freedom that does so. Freedom refuses to obey and this causes ego religion to change.

Ego can be found in many religions but is more predominant in Western religion as opposed to Eastern religion because generally, the Western mind contains more boundaries and more egos than the Eastern mind. Jesus said, "The scribes and Pharisees are like a whited sepulcher, all clean and shiny without, but within full of dead men's bones and all corruption."

Such is the case in all ego religions to a degree. They will offer you white shirts and ties, pageantry, and large and spacious buildings. To the keepers of Truth, who eat the delicious fruit of the Tree of Life, they will say, "I don't know you." Jesus said, "Not everyone that says unto me, Lord, Lord, shall enter into the kingdom of heaven; but he that doth the will of my Father which is in heaven."

Who is doing the will of the Father? Is it the ego, which is full of boundary and separation, which builds an empire of control, and claims to achieve union with the Father sometime in the future, but not now? Or is it the keepers of Truth, who transcended the ego, removed the boundary, promoted freedom, and experienced great joy in the present union with the Father now?

Keepers of Truth see the ego because they have transcended the ego, but the ego doesn't see Truth. The ego says, "Lord, Lord, have we not prophesied in thy name? And in thy name have cast out devils? And in thy name done many wonderful works?" If you are faithful to a religion, then look within and ask yourself what you see.

Do you see someone who says, Lord, Lord, claiming to have the truth, but making incorrect assumptions about what other faiths believe? Do you see someone who says Lord, Lord, dismissing other faiths as irrelevant until they become competition, wherefore they are content to see those blasphemers cast down to hell? Do you see someone who says Lord, Lord, citing good works as conclusive evidence of the preeminent righteousness of their religion?

It is the ego that cites good works as conclusive evidence of the tribe's ability to ensure its survival. It is the ego that sees other tribes as irrelevant to its survival, but if they are in competition with you, then the ego is content to see the other tribes destroyed. It is the ego that claims to have the truth but is often mistaken as it operates in a false reality. It is the ego that will say, "I don't know you," and in extreme cases, it will say, "I must destroy you."

The ego fosters enemies and is blind to Truth, but will project itself onto Truth and say that Truth has enemies. Jesus sees the ego because he represents a transcendent bridge to Truth. The ego doesn't cross the bridge. It projects itself onto Jesus and makes him into an egotistic Jesus who has enemies. It is this egotistic Jesus who will say, "I don't know you," and in extreme cases, this egotistic Jesus will say, "I must destroy you." Thus, the ego religion, chuck-full of dead men's bones projects itself outward to make an egotistic Jesus who must destroy, creating an even bigger pile of dead men's bones.

Awareness serves to remove the ego; likewise, awareness serves to remove ego religion. Accordingly, you should be aware of how thought and behavior patterns of the ego are manifested in the teachings and practices of ego religion. Some examples are as follows:

1. The ego is not a friend of the present moment. Therefore, ego religion will offer the promise of a future happiness that is separate from the transcendent now.
2. The ego desires control. Therefore, ego religion may offer you a controlling priesthood. It may offer tokens and signs to give you control over the angels of heaven to allow you entry into the presence of God. It may offer control of the dead through vicarious rituals of salvation for the dead. It may offer control over a multitude of wives, either in this Life or in a future paradise, with the number of wives being more than enough to start a women's softball league. Control arises out of identification with physical forms; therefore, control itself

is an idol. The ego does not realize that control is an idol, as it surrenders itself and worships the control of the ego religion. In exchange for this worship, the ego religion offers you the safety of the herd to protect you from the demon. But the ego is deceived, because as it worships the idol of control, it really worships the demon.

3. The ego craves security. Therefore, ego religion will offer you idols that are objects of security. Idols may be things such as paintings, statues, or the ink, glue, and fiber of a book. Idols may be people, who are held up as prophets, or authority figures, that must be obeyed. Idols may be a temple or exclusionary place, where the believers can securely gather to affirm their righteous standing.

4. The ego desires to be more than others. Therefore, ego religion may offer you the status of elite righteousness and provide you with a class of people that you can feel superior to, such as the non-believers, gentiles, infidels, or more generic evildoers.

5. The ego desires to have more than others. Therefore, ego religion may offer you a paradise that is complete with your kingdoms, your principalities, your mansions, your women, your status, and a cold beverage to sip on, as you watch the unrighteous burn in hell.

6. The ego desires to have more for the least amount of effort. Therefore, ego religion may offer you humility as a way to achieve elite righteousness with the least amount of effort. It may even offer you a deathbed conversion, the way with the least amount of effort of all.

7. The ego believes in guilt. Therefore, ego religion will assure you that you are not worthy of union with the Father and that your place is to confess your sins and ask for mercy so that you can escape the punishment you deserve.

8. The ego believes in loss. Therefore, ego religion will offer you a God that has only so much salvation to go around, but if you are faithful to the ego religion, then you just might make it someday—you just never know.

9. The ego has no relationships, only alliances. Therefore, ego religion will offer you an alliance with a God that will agree not to destroy you, but only if you pay and obey. And don't forget to kneel and place your forehead on the carpet, because this God is not your friend; you might make him angry with your lack of self-diminishment.

10. The ego is a false reality. Therefore, ego religion will claim that it has withstood all scrutiny. In reality, Truth withstands all scrutiny, not ego religion.

11. The ego does not realize that it sees a reflection of itself. Therefore, ego religion will claim that God must be worshipped. In reality, only egos wish to be worshiped. It will claim that others are deceived by a cunning devil. In reality, it is the cunning devil within that deceives the ego religion.

12. The ego has enemies. Therefore, ego religion will claim that all other religions are false and that the rites and sacraments of other religions are invalid and of null effect.

13. The ego builds an empire. Therefore ego religion will show you kindness, just don't ever leave the religion or abandon your oar, you will discover that the true message behind the smiles is, "We keep you alive to serve this ship. Row well and live."

14. The ego is not transcendent. Therefore ego religion never speaks the language of transcendence. It speaks the language of safety, certainty, and honor.

From the above examples of how the ego is manifested in ego religion, it becomes clear that many people may be pursuing a spiritual path that has run off the rails. These people should recognize the ego patterns both in themselves and in the ego religion to which they are identified because the ego is pathological and will be destructive even as it claims to be righteous above all.

It is no longer a matter of live and let live, as ego religion will say in its defense, affirming its right to exist, because ego religion, at its core, does not affirm your right to exist. It will claim that it does, but that is dishonest. Ego religion will also claim that it respects your beliefs; in turn, you must respect theirs. This is also dishonest because ego religion sees your beliefs as an abomination, and considers you worthy of destruction for harboring such beliefs. Therefore, to them, you do not have a right to exist.

Ego religion will say, "If we sit together and talk then we will find that we like each other more than we thought." This is true, but what is causing the animosity in the first place? It is a bounded belief system that calls for the domination and destruction of conflicting belief systems. It is impossible to have the unity that many religions say they desire because they will claim that unity can only exist under their flag. The only way there can be unity is when all roads lead to you and all roads lead to everyone else. In this way, all reference frames are likewise connected, and everyone is free to move in all directions. Unity is beautifully emblemized on the flag of the United States where there is a field of many stars and each one shines as brightly as the next.

Freedom is coming to this planet in a form that ego religion will regard as the unexpected because freedom is always painful for those rejecting freedom. For freedom to be, the ego must die, and that is why the age of the ego is ending. Likewise, the age of ego religion is ending because, as we can see, it is just a macrocosm of the ego.

Like McClellan, ego religion does a wonderful job of endearing and mobilizing a multitude of followers, as they declare themselves to be the theological saviors of the world, but inwardly, they scorn those whom they view to be their ecclesiastical and spiritual inferiors. Publicly, they are diplomatic, but inwardly they regard those on other spiritual paths to be idiots and baboons. They are disrespectful to Life, sharing a story of woe as they greatly overestimate the number of cantankerous devils lurking beyond the chapel door. They present a spit-and-polish front, as they remain in the safety of the village, where they claim to prepare people for the dimension of Truth, but are afraid to enter its depths.

CHAPTER 18

Lunacy

The accomplished author and journalist Christopher Hitchens is very critical of Christianity and religion in general. He believes that many of the fundamental teachings of Christianity are physically and morally destructive to society because they diminish personal responsibility. These teachings include vicarious sacrifice, taking no thought for tomorrow, and the forgiveness of sin, particularly without regard for the other parties involved, those considered victimized by the sin in question. Anti-theists such as Hitchens are not alone in this criticism because even the famous novelist and Christian apologist C. S. Lewis has stated that from the teachings of Jesus, "We must conclude that Jesus was either a lunatic or the Son of God, whose laws are broken and whose love is wounded in every sin."

I agree with Hitchens on vicarious sacrifice because vicarious sacrifice does not exist. There is no sin, only pain, and pain cannot be removed by sacrifice except possibly your own. Pain is removed in various ways, but a highly effective method is honesty. Grace operates in your Life to remove pain, but the pain is a false reality. False reality and dishonesty are very much the same thing. Therefore, Grace operates in your life to bring you to honesty where there is no sacrifice, only service. However, the presence of Consciousness within the field of lesser human consciousness has a powerful effect. It can bring about a transmutation of human consciousness and thereby remove pain and negativity. Consciousness is not broken and wounded by sin because sin is only an illusion.

I do not agree with him on the issue of "take no thought for tomorrow." This teaching is also called living in the present moment, or living in the now, which has already been discussed. Now is when you paint the ceiling of the Sistine Chapel and say you will make an end when you are finished. Now is when you build a Spacecraft named Eagle to land on the Moon before the decade is out. Now is when you make a deposit in your investment portfolio. Now is when you believe that a solution can be found.

Procrastination takes thought for tomorrow and worries about it so much that it doesn't take any action today. Righteousness takes thought for tomorrow and calls it a place where most people are relegated to oblivion. Thus "take no thought for tomorrow" offers the much-wanted benefits of initiative and empowered action, but also serves to eradicate the fearful and controlling judgments of righteousness.

I do not agree with him on the forgiveness of sin. Jesus told the sinners that they were forgiven and to sin no more. In other words, he told them to take responsibility and change for the better. Jesus did not consult with the other parties involved because they were to forgive as well. They are to forgive because in Eternity there is no such thing as guilt or loss. There is no guilt, only pain, the same pain that we should all recognize and help this world to overcome.

There is no loss because Eternity is infinite, it cannot be diminished, nor ever a victim. When you lay victimhood aside, you take responsibility. Jesus also gave us the Parable of the Talents, which suggested that we are expected to do something productive and meaningful with our lives. More importantly, Jesus spoke of being born again, which is the awakened state, where a person takes responsibility for themselves and this world by fulfilling their Life's purpose. A purpose is always an act of service that is productive and meaningful. The requirements of a healthy society are therefore preserved.

You transcend from being good to great through team effort, passion, and service. There is no such thing as blame in a team effort, passion, and service because there is only acceptance of the present problem for what it is, and finding a solution. On a team, the more you trust and empower others, the more power you have. True power is therefore found in trust and unity. Who is on your team? If you expand your perspective wide enough you will see that it includes planet Earth. If you expand your perspective high enough, who will see that it is the Cosmos? If you and the Cosmos are on the same team, then there is no one left to blame, there remains only great passion and great service.

Christopher Hitchens can be viewed as an archetypical person of keen intellect who observes how religion makes a mess of things, and he is seemingly left with no rational choice other than to toss the baby Jesus out with the bath water. This book is here to solve the problem of separation and this requires reconciliation of all people, for all is Truth.

It is amazing how many problems are solved when sin is removed from the table and replaced with the abundance of Eternity. Vicarious sacrifice disappears, responsibility is preserved, freedom shines bright, and the extension of Consciousness is maintained. Jesus ceases to be a lunatic. Hopefully, those with a keen intellect can now begin to reconcile with the baby Jesus.

CHAPTER 19

Duality

A theists object to the concept of God, describing a God that is always watching you, always controlling you, and always forcing you to obey. What they are describing is the ego, not God. The ego professes a false belief in independence that causes suffering. A true belief in independence recognizes that there is both dependence and freedom.

When you realize that what you depend on also desires your freedom, there is no contradiction between these two. I have been taught that there is no such thing as independent action, but there is imagination or free will. As long as you have free will, you are free to explore Life on your own terms without having to fit your beliefs within the stockade of conformity.

Conformity can turn your psyche into mush, and I am a case in point. I graduated from the University of Utah with honors, so I was no dummy. Yet at my first engineering job, I told my office mates that creation was very much a singular event and that evolution played no part in it. I said that the fossil record was being misinterpreted because the humanoid fossils were really diseased, and therefore misshapen human beings. I said that fake dinosaur fossils were placed in the ground by God to deceive us and test our faith in the Bible. This is what I was taught in my high school seminary class. My office mates didn't know whether to laugh at me or pity me. I was astute and able to deduce and abstract, but I had a divided, inconsistent reality. Some areas of my psyche had certainly been turned to mush by conforming to a Western religion.

You have your own free will, your own choices, and your own power to create. You are free to create, but you don't have free reign to destroy. That is where the Law of Attraction runs its interference. This Law says that imperfections that destroy will also destroy themselves. These imperfections are the ego that sees itself as an individual and not a duality.

Duality is the presence of separate essences interrelated in a control loop. A control loop may provide stability, but it may also conduct a controlled change. Creation is change; therefore, dualities themselves change. In creation, new dualities arise and others disappear.

In this human dimension, all things exist in a balanced duality or imbalanced duality. Balanced duality allows things to be. For a thing to be, the balance is not necessarily a perfect balance. If it is a perfect balance, then the push-pull nature of the control loop disappears and the duality disappears. If the duality disappears, then the thing ceases to be. In a balanced duality, there remains variation, choice, and freedom. This allows you the freedom to choose a trip to the mountains or a trip to the beach. An imbalanced duality allows creation to operate at full throttle and allows new dualities to arise. Often when creation operates at full throttle, it is perceived as the unexpected.

To better understand duality and control loops, we'll share a story of a chef, a kitchen, and a stove.

When the stove is off at night, it exists in a state of perfect balance with the kitchen. If the kitchen temperature decreases during the night, the temperature of the stove will also decrease. It will remain balanced with the temperature of the kitchen. The kitchen and the stove do not exist in a duality at night. They are perfectly balanced and simply exist together as a kitchen.

If the stove is turned on, the temperature of the stove will increase by hundreds of degrees. This will also increase the temperature of the kitchen. However, the kitchen will not increase to match the temperature of the stove. This is because the kitchen, extending into the restaurant and the outside environment, is much more vast than the stove.

The chef arrives in the morning to begin his day in the kitchen. He turns on the stove by turning the knob to 300 degrees and disrupts the perfect balance of the kitchen and stove. A new duality is created, the duality of stove temperature and kitchen temperature. The stove heats things up, but the kitchen cools things down. As the stove heats up it obtains a stable temperature of 300 degrees.

Within the stove is another duality. This duality is a control loop between the stove's heat source and temperature sensor. When the temperature sensor detects that the temperature is too low, then the heat source is increased. When the temperature is too high, then the heat source is decreased. In this manner, its temperature remains stable.

The chef places his hand close to the surface of the stove and feels that it is heating up. The chef checks the temperature gauge on the stove, which shows that it has reached the desired temperature. The chef and the stove now exist in a control loop. If the gauge is on, the chef will cook, as he cooks, he continues to check the light and feel for heat from the stove.

The chef receives an order of eggs. He places eggs on the stove. The chef and the eggs now exist in a control loop. He observes the eggs; when they appear ready to turn or to be removed, he does so with a spatula. Thus, the eggs control the chef and the chef controls the eggs.

At the end of the day the stove is turned off, cleaned, and polished; again, it cools down to the temperature of the kitchen. The duality of the stove temperature and the kitchen temperature disappears.

Note that both the kitchen and the chef controlled the stove at different times. The chef controls the temperature of the stove when he is present. The kitchen controls the temperature of the stove when he is absent.

Within the stove is the control loop of the heat source and the temperature sensor. This temperature control loop within the stove represents intelligence. It senses, reacts, and serves the purpose of creation. The chef also represents intelligence. The kitchen provides both the stove and the chef with electricity, natural gas, utensils, refrigerated eggs, and a sheltered environment; it all arises from intelligence.

If you were a character in this story of the chef, the kitchen, and the stove, which character would you be? Most people might consider themselves to be the chef. However, if you are an ego, then you are the stove.

The stove sleeps at night and is controlled by the environment. Morning comes and the stove finds its temperature turned up to serve the purpose of creation. It has an internal control system preventing it from becoming too hot or too cold. Eggs appear on its surface. It applies its heat to transform the eggs. A spatula appears and flips the eggs and the stove continues to apply its heat. Soon a spatula appears and removes the eggs. During the course of the day, the stove's surface is scraped, cleaned, and polished, therefore, it is never quite the same as it was the day before. At the end of the day, the stove finds its temperature being decreased again. It tells itself that it did a fine job that day as a fryer of eggs, as it powers down for the night.

You sleep at night and are controlled by the environment of your house, your bed, and your blankets. You awake in the morning. Your body becomes operational to serve the purpose of creation. You have an internal control system that prevents your body, and hopefully your emotions, from becoming too hot and too cold. You perform tasks at work or at home. People and objects appear before you. You perceive them, communicate with them, and act upon them. People and objects are transformed. The people and objects that share your day are eventually removed from your presence. You are transformed by this day because you are not quite the same as you were the day before. You relax and retire to bed. You tell yourself that you have done a fine job at your tasks that day, as you drift off to sleep.

An ego believes it is separate from the environment and other people. It believes it is independent in thought and action. The ego does not perceive that it exists in duality with creation, that creation is controlled by Truth.

Even if the ego did recognize the duality, it would likely claim an equal share of control in the duality. However, the ego is the stove, which is controlled by the kitchen and the chef, and has very little control as it serves the purpose of creation. The purpose of your creation is for you to realize that you are not an ego but are instead a higher self. The higher self is the chef and is also the kitchen. You don't see the higher self because it lives in the dimension of Truth, but you perceive its effects in the human dimension.

The ego sees itself as an individual facing a world of opposition, but the ego is in a duality with the environment. The ego sees the environment as a threat, but the environment gives gifts that allow you to survive, and as you survive you serve creation. As you serve creation, you change and exist in an ever-changing duality with the environment where you change the environment, but the environment changes you. As you change the environment, you may choose to kill or destroy. If you kill and destroy too much and become a threat to the environment, the environment may in turn threaten you.

When the environment has become a threat, your choices have been greatly limited. You may find yourself trapped inside the bunker, with the thunder of the Soviet Army bearing down on you, sitting with a cyanide pill in one hand and a pistol in the other, as Hitler did. The only choice you have at that moment is the method of suicide. You may find yourself in a bunker of a bounded belief system with the thunder of the unexpected bearing down on you as you sit with your closed canon of scripture in one hand and your list of liars and blasphemers in the other. The only choice at that moment is dogma or change.

You have intelligence and life. You are born into a duality with the environment. Likewise, the environment has intelligence and life. It exists in a duality with you. You are therefore Life in a duality with Life. This duality exists to guide you to the Truth so that you might explore the depth of the Truth as a human. It does not exist so that you can remain in the bunker of a bounded belief system.

Religion does not recognize duality and therefore offers you a bounded belief system that separates you from Life. This cannot be denied. Religion offers you a future paradise and thereby perpetuates the separation. If you are separate from Life, then Life remains your enemy, and you become death. If you become death, you are blind to Life; you cannot conceive the Truth that is ever-present with you and within you.

Truth is right in front of you. Truth is all around you. Truth is the least of these whom you don't forgive and instead condemn. Truth is the blasphemer who tells you that he has fulfilled the Law and will not obey you. Truth is the freedom that demands all people to be free. Truth is the one who tells you to change. Deny change and you cease to explore. Cease to explore, and deny the purpose of your Life.

In Truth, there are many dualities. Consider the dualities of friendship, marriage, religion, and culture. There is duality when you feel less than and desire more. This duality is unstable because the desire for more is not the opposite of being less. This unstable duality is painful. It can be very destructive, but it ultimately perfects us and drives the creation process. It forces the mind to find the Truth within. The Truth within is a stable duality of being less and being more, but being less is not to be confused with humility, and being more is not to be confused with arrogance.

In Eternity, you discover the duality of the large and the small. You must be both large and small to step into Eternity. You must be a mighty king. You must also be the least of these.

CHAPTER 20

Freedom

Jesus said, "I have come to preach deliverance to the captives and set at liberty them that are bruised." If you are not free, then you are a captive and you are likely bruised by your burdens. He also said, "Come unto me for I am meek and my burdens are light." There is nothing quite as meek as freedom.

Freedom does not demand obedience. Freedom tells you to follow your own muse and bliss. When you are free to do so, your burdens become very light indeed, and the powers of Heaven overtly assist you in accomplishing your Life's purpose.

The solution to the problem of the human dimension is freedom. Freedom is more than the solution; it is the single concept that best describes the nature of Truth. Freedom cannot be contained; it must expand and create. Freedom is always shared and contains no conflict.

When you are free, you are in harmony with Life, and when you are in harmony with Life, you are free. When you are free, you are free to share and create. This is why free nations live in relative peace with one another and become wealth-springs of creativity.

Freedom is synonymous with possibilities. If all things are possible, then freedom becomes complete and absolute.

Eternity is freedom that is complete and absolute.

You experience Eternity when you awaken to Truth.

Therefore, Truth is your freedom that is complete and absolute.

When you receive Truth, you live in a state of Being. In Being, your free will merges with the free will of Consciousness. You are free because you become Consciousness. Consciousness is perfection. Perfection is freedom.

Freedom is sharing and sharing is service. Service is creation and creation is service. Creation is ever beautiful. Beauty is in the eye of the beholder and is therefore self-indulgent. Creation is therefore a beautiful and self-indulgent service.

We are taught to serve one another, that we are uplifted in the service of our fellow man. The concept of service being self-indulgent is therefore nothing new. We are also taught that service must be done willingly. The concept of self-indulgent service, being willingly done, is also nothing new.

However, the concept of creation being free and self-indulgent may be something new. If creation was anything less than free and self-indulgent, then creation would find something better to do.

Some people equate freedom with excess, paralysis of choice, and dissatisfaction. Paralysis of choice comes from too many options causing an inability to choose. It creates dissatisfaction because even after you make a choice you wonder if it was correct. What they are describing can be called the grass-is-always-greener-on-the-other-side syndrome.

This syndrome is not the result of freedom but is a manifestation of the divided mind, which is the bottomless pit that can never be satisfied. The divided mind lives in fear; it has become addicted to belief in loss. The divided mind practices control, in an attempt to overcome loss, but control never satisfies because the belief in loss is never overcome. It is never overcome because the divided mind places a boundary between itself and Truth where there is no loss. There is only creation.

Freedom is not self-indulgent individualism, hedonism, or minimalism. I am free, and I express hard work, excellence, honesty, forgiveness, generosity, service, intelligence, laughter, love, and appreciation. I support a healthy lifestyle. I support marriage. I strongly support the unborn. But mostly, I support freedom. I am freedom because my life is mine and mine alone, so is yours. Moreover, I view freedom as the solution to problems such as an unhealthy lifestyle, a difficult marriage, or a choice between life and death because in freedom there is no judgment.

People are constantly judged to be less. They therefore live in constant fear of not being enough or not having enough. Their decisions are formed from a perspective of fear and scarcity, rather than a perspective of harmony and abundance. They can suffer a paralysis of fear and become incapable of making any decision at all. Through fear, their life remains limited to a place of bitter disappointments where they can find no harmony or abundance.

If there is no judgment, then fear is removed. If fear is removed, then the limitation is removed. When the fear limitation is removed, there is freedom. In other words, freedom perpetuates freedom and cannot be contained. By the Law of Attraction, you can only see yourself. If you see a forgiven world, then you are forgiven. If you see a sinful world, then you are the sinner.

There can be no boundary between you and the sin you perceive; therefore, the sin you perceive is in you. Furthermore, there can be no boundary between sin and no sin. If there is sin, then God is sinful.

There is no such thing as sin, but there is fear limitation. You might say that there must be no boundary between fear limitation and no fear limitation. This would mean that God must be fear limitation. However, this is not the case because fear limitation is a duality; it is only relative truth.

Some common fears include fear of public speaking, fear of crowds, and fear of death. But perhaps the worst fear of all may be the fear of failure, which makes it impossible for your dreams to become reality.

To examine the concept of limitation, imagine a person whom you would claim is the most sinful or evil person in history. Now answer the following questions regarding this person. Would you trust this person to live in your home? Would you trust this person to babysit your small children? Would you trust this person to educate your older children? Would you trust this person with your bank accounts? Would you trust this person to run your company? Would you trust this person to run your country? Would you trust this person to safeguard your freedom? I assume all of your answers are no.

Now imagine a person whom you would claim is the best or most virtuous person in history. Now answer the following questions regarding this person. Would you trust this person to live in your home? Would you trust this person to babysit your small children? Would you trust this person to educate your older children? Would you trust this person with your bank accounts? Would you trust this person to run your company? Would you trust this person to run your country? Would you trust this person to safeguard your freedom? I assume all of your answers are yes.

If you are sinful or evil, then there are many things that you cannot be trusted to do. Your options are therefore reduced. In other words, you are limited.

If you are fearful, then there are many things that you will not attempt to do. You do not learn nor progress, and therefore you remain limited.

If you are a bundle of fear limitations, then your options are reduced. If your options are reduced, then you are not free. If you are not free, then you don't have Truth and are stuck in an illusion where a divided mind attempts to control others.

The reality that sin does not exist may come as a surprise, but hopefully, it comes as a relief. Many people seek atonement, which is the removal of sin. The divided mind places sin as absolute truth because it places sin within the context of absolute Law. It sees absolute Law because it sees absolute loss. There is no such thing as loss. There is only perfect Love, and perfect Love always forgives. In fact, the concept of forgiveness does not exist in perfect Love. In perfect Love, there is only acceptance and sharing that extends to you. As it extends to you, it lifts you above fear limitation and you become perfect Love.

To be lifted above fear limitation, you should believe in the impossible. When you believe in the impossible, you believe that it is possible for your dreams to come true. You can also believe that it is possible for you to be deceived.

Awakening is a gift of Grace. There is no fixed set of rules or patterns of behaviors that bring you this gift. Therefore, no one can lead you to an awakening event; they can only serve as guides. Accordingly, various guidelines can be offered to remove ignorance and increase your understanding of the human dimension.

Your understanding of the human dimension is a reflection of yourself. Therefore, for your understanding to increase, you must increasingly know yourself. However, if you attempt to explain who you are, then you are telling a story that is nothing more than a collection of thoughts, and thoughts are only relative truth. Therefore, your attempts at storytelling only demonstrate that you are someone who accepts relative truth as absolute truth and is thereby deceived. To know yourself is not to tell a story of who you are; it is to understand what you are not.

Understanding grows from the following simple principle: Everything is connected. As you remove boundaries, you will begin to see these connections, and you will be inspired to remove more boundaries. There can be no boundaries between people. Therefore, you must be open to radically changing your concept of individuality. At some point, you may be inspired to remove the boundary that separates your thoughts from reality, but to do so, you must first find it.

CHAPTER 21

Obedience

M any run on the treadmill of obedience, calling it the pathway to heaven; they will say, "Spiritual development requires a lot of spiritual work because life is a test to determine what you want. You must be determined to follow Jesus at all cost, then eventually the yoke of your burdens will be made light."

The paradox of the treadmill is this: If life is a test to determine what you want, then why are you choosing a test?

The paradox places you in conflict because you never arrive where you want to be. All you want is to be in a situation where you're not constantly tested. You try to tell yourself that the endless testing builds character, but all this does is confirm that your character is never good enough.

There is a contradiction of cosmic proportions when you associate following Jesus with a cost because the cost is synonymous with loss. Who is it that believes in loss? Who is that believes in guilt and says you are never good enough? You already know the answer to these questions from reading this book.

People who say life is nothing but hard work are called doers. They are egos on the treadmill of obedience. They can only imagine future happiness and never make friends with the present moment. Happiness is not about doing; it is about enlightenment because enlightened doing is doing precisely what you want to do. You no longer consider it work because you are one with the Cosmos' creative power. It continuously flows through you and brings you happiness, regardless of what you are doing.

This is Being.

A Dewar is a container that keeps things cold. A doer has a heart kept cold because life remains a struggle. It remains a struggle because they lack coherency between their actions and their desires. Coherency is lacking because they are not honest, and they are not honest because they are deceived. They will tell you that God gives you a test that can never let up because it is the struggle that makes you pure, yet Jesus was sent to relieve you of at least part of your struggle. Which is it? Is it a struggle that can never let up, or is it a struggle that is at least partially relieved? This is another example of the contradictions and conflicts of the divided mind.

There once was a divided mind that was a happy and playful little boy until age ten when his younger brother died of measles. This caused him pain and changed him, and he began to build his wall. He wished to be an artist, but his father forced him to attend technical school instead. His father struggled to earn a living and abused him until age thirteen when he eventually died. He expressed gross disrespect for his school amongst his friends and one of his friends informed the schoolmaster. For his disrespect, he received a dressing-down from the schoolmaster that reduced him to shivering jelly and he was expelled. He never returned to school again.

At age fifteen, he became a Christian. He applied to art school twice, but he was twice rejected labeled unfit for painting, and was advised that his talent was in architecture. He then was determined to be an architect, somehow Soon his mother died, and he was essentially on his own at age eighteen. He pursued architecture, but at age twenty, he found himself broke and living in a homeless shelter. This divided mind had struggles, but he came to believe that the struggle makes you pure. In fact, he wrote a book about it called *My Struggle*. The book is more commonly known as *Mien Kampf*. This divided mind's name is Adolf Hitler, and his beliefs caused him to unleash a monster of global war and genocide against what he considered the opposition.

Your belief in opposition and purifying struggles may not start a global war, but you should recognize that your belief system shares the same thought structure as someone who *did* start a global war. The question to ask is, how can a thought structure that leads to global war also lead to Heaven? The answer is that it can't. It can only lead you to give up the struggle. Then your eyes will be opened and you will see what Jesus meant by burdens being light.

I once saw Life as a struggle, but I also wished to understand my reality. In *The Secret*, the Cosmos is modeled as a genie who says, "Your wish is my command." This is another way of stating the Law of Attraction. I followed Jesus, and he led me to the Genie. I now live in a game of make-believe where the Genie says, "Your wish is my command." I now understand my reality. I am the Cosmos. I am happiness, laughter, and sharing. The separation has been undone and Life is no longer a struggle. Life is now Oneness and Truth.

Truth is mainly hidden from the human dimension, but like all things hidden, it is here. From just beyond human perception, Truth is here. It places images and thoughts in your mind. It causes you to ask questions and then gives you answers. Throughout Life, Truth looks at you with a playful grin and whispers, "Do you not see me?"

You are drawn to the illusions of this world and become possessed by a demon within that both loves and hates the illusion. You see illusions as the solution to your endless wanting, but this is the cause of all human suffering. You can wish that the illusion had never come to you and even desire that the demon be killed, but the demon within has a part to play in the destruction of the illusion because it is what drives you to Truth. Truth places you here to learn the lessons of Grace, and to this end, you are meant to have the illusion and suffer the demon.

A young Islamic man who is concerned about the injustice he sees perpetrated by Jews and the West, will ask his Imam, "What do I do?" The peaceful Imams will say, "Be confident that good will win over evil." This may sound like a message of peace, but what image does this leave in the young man's mind? Who is the good and who is the evil?

The implication is that Islam is good, and the Jews and the West are evil. This gives the ego and pain body more combustible fuel for opposition. It increases volatility even though the Imam may sincerely believe that he has given counsel that promotes peace. It is not the desire for good to win over evil that brings peace. Peace comes when peace is accepted as the present reality.

The demon exists in human consciousness and also exists on a global scale. It desires revenge and will use any tool available to achieve this end, including crashing civilian aircraft into the Twin Towers of New York City where thousands met a sudden and brutal end to their lives. These attacks were painful, but all pain is part of a landscape of possibilities that allows solutions to be found.

Members of New York's finest, including policemen and firefighters, suffered death in the attack and some of the survivors were heard to say, "Osama bin Laden can kiss my Irish ass." This sentiment was widespread in an unprecedented bipartisan fashion and the Western ego responded to the unexpected with war, rather than acceptance. As a result, many tens of thousands *more*, including tens of thousands of Americans, have been wounded, maimed, or killed. If acceptance had been chosen, then the violence of war would have been avoided, and the trillions spent on war could have instead been used to rebuild the Twin Towers and transform the slums of Gaza, the West Bank, and Lebanon into a paradise of the Fertile Crescent.

When you are enlightened, you will not respond with war when someone attacks the Twin Towers of ego and pain body. If you say more was attacked on 9/11 than ego and pain body, then think again. They *are* your pain, so what else could be attacked other than your pain? Jesus told you not to respond to an attack with another attack. He didn't teach this to diminish you. He taught this so people would open their eyes and see that all people are something much more than the fearful little me that is lost in dualistic illusion.

I have toured much of Israel, including Jerusalem, Galilee, Golan Heights, and Masada. I had an interesting experience on our tour bus when we picked up a hitchhiking Israeli soldier. Israeli soldiers always travel with their weapons, and buses and other civilian vehicles are expected to stop and provide them with transportation. This soldier sat next to me with his Israeli-designed and manufactured Galil assault rifle.

We talked about the rifle rekindling my interest in military hardware. Another tourist from England who sat on the bench in front of us turned around and started talking to the soldier as well. He asked the Jewish soldier a couple of questions and then said, "You know, you have to accept Jesus Christ as your savior." The Christian didn't seem to appreciate the context of his laconic statement. The young soldier is required to carry his weapon with him. He lives in a land that has seen centuries of warfare brought about by religious zealots claiming the authority to dictate the actions of others. The soldier holding a loaded weapon is a guardian of the freedom that allows people like the Christian to make such authoritative statements without being killed by another religious zealot of an opposing view. The soldier replied, "Yeah, whatever," and then he closed his eyes to rest, as he kept both hands on his rifle.

Islam demands obedience. Christianity demands obedience. Many humans demand obedience. It is time for freedom. Obedience must end. The more people hesitate to do so, the more opportunity the demon has to make them suffer and take the lives of thousands of people, who simply went to work on a Tuesday morning.

The Cosmos is full of happiness, laughter, and sharing. The Cosmos wishes to share its laughter and happiness with you. The Cosmos is freedom, so you must choose the reality that you desire. If you can't choose the reality that you desire, then how could you ever be happy?

Doers stand behind the veil of a controlling belief system that fails to recognize the transcendent Jesus. I was a doer but no longer. I invite all doers to Be. That is, awaken to the happiness, laughter, and sharing that is your home. Buddha tells you Nirvana is here.

Jesus tells you that Heaven is here. Who tells you that it is not here and robs you of your happiness?

Is it your neighbor? The television? Your smartphone? Your religious authority?

Whoever it is, it is a person stuck in the human dimension and does not wish for you to make your escape.

CHAPTER 22

Jerusalem

The empire of the West is a grain of sand in the vastness of the Cosmos. The empire of Islam is also a grain of sand in the vastness of the Cosmos. However, together these two empires form a duality that created the sands of the sea. Jerusalem is a balance point in the duality of Islam and the West that can express a false reality of violence or a coherency of peace.

Truth has reestablished Israel and blessed it with fantastic victories. It stands as a promise to Abraham and Jacob for a purpose. Anyone who wishes to harm Israel brings great pain upon themselves and this will always be the case because Israel shall fulfill its purpose.

Israel exists in the Loving arms of Truth, but so does everyone else. Within Israel, there are Jews who see much separation because of the pain of the past. The more pain of the past you carry the more separation you will see. In their pain, they oppress Muslims and Christians. They choose the path of conflict and the ego mirror reflects back to them a world of conflict. Anyone who seeks conflict in the ego mirror experiences more pain.

The Torah teaches that Abraham was promised that through him and his descendants, all the Earth shall be blessed. Jews bless the Earth. Jesus blesses the Earth. Muhammad blesses the Earth. The message of Muhammad directly created the Islamic empire, which formed a duality with the Christian empire and this duality drove the engine of creation to the present day.

The creation of this empire was the mission of Muhammad. He was a prophet, but he is not the last, and the issue of who is the greatest is meaningless because it forces a judgment of relative truth and ignores a higher perspective. Islam claims that Muhammad supersedes Jesus, but from the perspective of peace, Muhammad's message is a lesser message than that of Jesus.

The message of Jesus and his presence in this world is the crown jewel of the promise made to Abraham. I can say this without making a judgment, because Jesus taught freedom, and freedom does not contain a judgment. The message of Muhammad is a message of obedience and therefore contains judgment. A message of judgment is a message of separation, which is the ego. Muhammad's message contains ego and it therefore builds empire. It cannot help but do so—such is the nature of ego. So when they talk of peace, they may feel sincere, but deep down they are lying; the ego is always lying. Then was the Christian commentator correct when he said Islam is lying about not having enemies? No, he was not correct because Islam doesn't lie. They, like Christianity and Judaism, contain ego. The ego believes it is completely honest because the ego is deceived.

Whether you regard Jesus as an awakened human, an Avatar, or the Father in the flesh, he is a powerful voice of transcendence because he remained true to the perfect Love expressed by Consciousness. He taught forgiveness and said you must Love your enemy. He never advised his followers to regain stolen wealth. He never chose to defend himself. He was not an ego and his message was not given egoic characteristics until after the crucifixion.

It may sound like I am setting one prophet above another after all, but I am not. I am setting one message above another. Setting one message above another does not set one prophet above another because a prophet is not his message. A prophet is not his message because by definition the message comes from God. Islam will tell you that Muhammad's message is greater because he is the greatest prophet. They thereby make a false association between a prophet and his message.

On the other hand, Jesus tells you, "It is the Father living in me that is doing this work." He thereby correctly disassociates himself from the message. This statement of Jesus can also be expressed as, "The Father is the dancer, and I am the dance." Jesus never missed a beat in this dance because he is Truth and Truth is given in its proper place. Sometimes the proper place is to unite the Arab people and expand a monotheistic empire. Sometimes the proper place is an expression of perfect Love so you can find the Truth within.

Some say that the meaning of the word Jerusalem is, "Abode of Peace." But it can also be interpreted to mean, "Teach Oneness." Throughout much of history, Jerusalem has been anything but an Abode of Peace, but it has always been here to teach Oneness. Truth uses Jerusalem to teach Oneness, but Truth also teaches Oneness in other places. Any place where Oneness is taught is a New Jerusalem. Anywhere there is a New Jerusalem you will find Truth.

Servants of Light are those who seek righteousness and wish to serve God, but they contain a component of the demon and see a world of good versus evil. In Jerusalem, or a New Jerusalem, Servants of Light teach that there is one God and that we must serve the one God.

In Jerusalem, there are Servants of Light that teach freedom and oneness, but they also contain a component of falseness. In their falseness, they also build separation and empire. In the land of Jerusalem, the freshwater is north of the salt water. This symbolizes that in Jerusalem there is God, and the Servants of Light pull him down to make him one with man. They thereby make God into an ego that is jealous and vengeful. In Jerusalem, there is the Messiah, but they pull him down to make him an ego. They look in the ego mirror and call the Messiah a violator of the Law and a blasphemer as they nail him to a cross. Then they sit with their hammers in their hands and await the Messiah. They claim that they are the Servants of Light and will strike with their hammers if anyone calls them servants of falseness. In summary, they make God an ego.

In New Jerusalem, there are Servants of Light that teach freedom and oneness but they also contain a component of falseness. In their falseness, they also build separation and empire. In the land of New Jerusalem, the salt water is north of the freshwater. This symbolizes that in New Jerusalem there is man, and the Servants of Light pull him down and enslave him to make him one with God. They thereby make man into an ego that is jealous and vengeful. In the New Jerusalem, there is the Messiah, but they pull him down to make him an ego. They look in the ego mirror and call the Messiah a violator of the Law and a blasphemer as they nail him to a cross. Then they sit with their hammers in their hands and await the Messiah. They claim that they are the Servants of Light and will strike with their hammers if anyone calls them servants of falseness. In summary, they enslave man to make him one with God, and they make God an ego.

Servants of Light serve the purpose of creation as they teach the message of Light, but they also fight wars of liberation against falseness. They also suffer the pain of the message, because the message is of the ego, which compelled them to fight wars in the first place. Servants of Light fight the falseness, the falseness in them.

The only act of Jesus that can be considered physically violent was cleansing the Temple, condemning the business activities of an ego religion. However, this conflict was perfect in its place because it led to the crucifixion. The cleansing of the Temple is also a message that stands as a warning today. The warning is that the age of empire is over, and deception for profit or control is no longer justifiable.

Mount Moriah in Jerusalem is said to be the place where Abraham went to sacrifice his son Isaac. It is also said to be the place where Mohamed transcended to high levels of Truth. It is also a place where Jesus taught oneness. Jerusalem is a Temple of sacrifice, transcendence, and oneness. Abraham and Muhammad have already symbolized the first two of these components on Mount Moriah. The third component will be fully symbolized when the children of Abraham are One.

Truth comes to remove falseness and cleanse the Temple. When the Temple is cleansed, then all is One. When all is One, the Earth shall inherit Truth.

A few decades after the death of Jesus, the Jews set aside his advice to live peacefully with the Romans and revolted. Consequently, the Romans destroyed Jerusalem and the Temple, and many Jews were crucified. Some Jews fled to the mountaintop palace fortress of Masada. The Romans surrounded Masada and patiently built a surprisingly large ramp to the top of the plateau. The building of such a large ramp was unexpected. As the ramp neared completion, the Jews within the bunker of Masada chose suicide.

Empires are of the demon and the demon is to be subdued. Those who insist on building an empire will find themselves on a Masada. They will be huddled in fear limitation, as the unexpected encircles them, aiming to take them from the bunker. They will then discover that their empire stands alone and is nothing but a grain of sand.

Who are the sands of the sea? They are many things, including the wisdom of the Jews, the Love of the Christians, and the purity of Muslims. They are the ones who will bless the whole Earth by subduing the demon. Then the children of Abraham will no longer sit with a hammer in their hand, and Jerusalem shall forever be the Abode of Peace.

CHAPTER 23

Grace

As we continue to rise above fear limitation, we learn Grace. Therefore, I describe the human dimension as a classroom of Grace.

The word Grace is open to interpretation. Most who believe in Grace would agree that Grace has at least the following three properties: It is the Truth that allows the universe to arise and exist, there is a separation between man and Grace that causes Grace to be at least partially hidden from view, and Grace seeks to undo the separation and unite with man. Allow this to be our common ground as we proceed.

Grace reveals the human dimension for what it is: a labyrinth of corridors, turns, and dead ends formed by a multitude of fences, the only escape from which is the narrow gate. The fences that form this crazy maze are the boundaries that exist in you. These boundaries are fear limitation because fear limitation places boundaries to make a limited reality that is more known, more familiar, and less fearful.

To rise above fear limitation and learn Grace, you must remove boundaries. To remove boundaries, they must first be recognized, but the trouble is, they are well hidden. However, you can recognize boundaries by their shadows, which are your thoughts and emotions that surround the separation.

The separation is the fall from Grace, but Grace is every human being. Therefore, the separation between you and Grace is equivalent to the separation between you and other people. This includes people who don't believe in your system.

The separation exists at both the personal level and the collective level. The collective level includes the groups, religions, and nations that you identify with. Whatever you identify with becomes part of you. If you identify with separation that exists anywhere at the collective level, then that separation also exists in you at the personal level.

Many people say they believe in Grace, yet they still perpetuate separation. These people are religious but prescribe to a bounded belief system. Their bounded belief system is a little box; they place God within the box and thereby limit God.

Fear places a boundary between that which is fearful and that which is feared. Therefore, they make a limited God that is to be feared, just a mirror image of their own fear limitation that seeks to control. Your beliefs are your perspective, and they can never be anything more than that. If you wish to remove separation, then your perspectives should be compatible with other perspectives that exist on Earth, or in Space.

When spiritual people place their burdens on Grace they relieve the weight of their burden, but does this diminish personal responsibility? If you are a member of a team, it is perfectly reasonable to assign a problem to the team member who is best equipped to solve the problem. This does not compromise personal responsibility because everyone on the team still provides their own contribution. It is also an act of trust that strengthens relationships. If you trust Grace, then your relationship with Grace will be strengthened. This is a very desirable relationship to have because Grace often solves the problem by informing you that it is not a problem to begin with.

Grace is freedom and freedom by definition must take full responsibility. Blaming others is a judgment that strips you of freedom. If you blame others, then your desired outcome is barred from fruition by external forces over which you have no control;

therefore, you are not free. Some blame the original sin of their ancestors. Some blame their parents. Some blame a political party. Some blame another religion.

If you wish to be free, then you must not blame nor judge Life. When you are free, you are Grace and there is no more sin, whether it is original or something plagiarized. Original sin can be defined as a false belief in independent thought. Interestingly enough, the term "original sin"—that has been used by Western religion for millennia, and has been given a great weight of seriousness—is a pun.

Grace does not remove responsibility; it brings you to the Truth. Grace is everything and if you don't wish for Grace to join your team, then you make a choice to deny Grace. We have the ability to make choices, but people are hypnotized by pain and therefore can never be held responsible. If you require people to be held responsible, you are actually expressing pain within yourself. You are saying that you can be diminished by their lack of responsibility. We must always remember that in Eternity there is no lack and you are not diminished by pain except the pain that exists in you. You have pain and may deny Grace, but Grace knows that you are not responsible for such misperceptions and therefore remains on your team.

Grace always brings you to Truth because there is no sin. If you believe in sin, there is blood on your hands. It is you who demands sacrifice. You cannot wash the blood off your hands with the blood of another. To wash your sins in the blood of the lamb is to kill the innocent to save the guilty. It is an affirmation of guilt and loss, which is of the ego that does not know Love. This is why tyrant egos kill the innocent to protect the guilty, but Love does not. Love is not an ego. It is blind to guilt and loss. Love only knows perfection.

What do you wish to see when you look upon the One Life? Do you wish to see an ego, or do you wish to see Love? Do you wish to see sin, or do you wish to see perfection? Do you wish to see enemies, or do you wish to see friends? Do you wish to see oblivion, or do you wish to see abundance?

It is the ego that sees sin, enemies, and oblivion. The ego also sees honor as the end result of righteous living. Truth is what you desire, not honor. Keepers of Truth aren't concerned with honor because they are complete. Like the flowing of the Tao, they love and nourish all things. They love and they serve. They wish to place you at the head of the banquet table. They are ridiculed by friends and family and throughout history some of them have been put to death, and yet they forgive and do not attack because they are the Love that sees the perfection in you.

Some people say that the teaching "Love your enemies" is grossly irresponsible because it leaves you vulnerable and forces you to compromise your principles. Vulnerability is an affirmation of the reality of loss, which is of the ego. Your principles may also be of the ego, and if they are, maybe they should not only be pruned, but completely rooted up.

Peaceful movements promote peace and are of Consciousness. Anti-war movements are of the ego and promote conflict. The ego doesn't want peace; it just wants to be against something. It wants to find something that it can label bad so that when judgment comes and casts its condescending gaze upon it, it can claim to be good.

The ego does not comprehend the concept of "Love your enemies" because that would force the ego out of opposition. If the ego doesn't have opposition, its false reality is demolished and the ego ceases to be. The ego therefore regards peace as suicide and insists that enemies be cultivated. "Love your enemies" does not mean surrender to them. As I have said, Consciousness serves but never obeys. It never obeys because obedience is not peace. Freedom is peace.

Creation is change and you are free to participate. You should view change as an opportunity for higher expression and experience, and not as an opportunity to escape the present moment. If you are friends with the present moment, then Life will work with you to create the change that you desire. The change you desire will allow you to escape the gates of hell, and the future will no longer resemble a painful past.

You will no longer be fettered by karma, just another form of pain. Anything that pulls up your anchor from the present moment is a storm of pain that you have not yet overcome. Earning your way into heaven is an example of such a storm. If you are engaged in earning enough credit to receive your heavenly diploma, then you are not building character, you're just reinforcing pain.

As you reinforce pain, you will say that you believe in love and peace, but you will also find that your Life remains woeful and that events happen where you must make excuses for outbursts and false assumptions. However, if you simply accept the gift of Grace, then you will find that your stress level is greatly reduced because you are at peace with the present moment.

Some Christians believe that they have accepted the gift of Grace but they have not. Instead, they have accepted the Arminian view that asserts you accept Grace by your own decision. If you say you accept Grace by your own decision, then this decision is made outside of the bounds of Grace. This places a boundary between Grace and no Grace and thereby separates them from Eternity. To fully accept Grace, you must accept that all is Grace, and stop taking credit for making the decision. I have heard many Christians tell a story of how they were led to Grace by Grace.

Eternity is a place of infinite possibilities and infinite perspectives. It is possible that your perspective must overlap, commingle, and merge with the perspectives of another planet. Many people are compatible with such a possibility because they are free-minded and open to change. Most importantly, they are not dedicated to a dogmatic set of beliefs regarding salvation.

If you are dedicated to such a belief system, or not sure if you are, perhaps you should ask yourself the following questions: When contact is made with another planet, will you engage with them in debates based on scriptural text? Will you tell them they must believe in your religion and engage in your religious rites to be saved? Will you tell them that they must believe in your prophet? Will you maintain that you are the elect of God and they are not? Will you threaten them with damnation?

Dogmatic belief systems presently interact with other perspectives on Earth with control, judgment, and damnation. This is intensified when someone within the belief system jumps ship and changes perspective. Under the most extreme dogmatic belief systems, you can't jump ship without walking the plank. Leaving the ship is grounds for execution.

Your perspective may not be compatible with mine, but mine is compatible with yours. Your perspective is one of an infinite number of perspectives I am compatible with because I am Truth. Truth has no opposite and does not condemn because it has no enemies, only friends. Truth is your friend and is operating in your Life to awaken you to its presence and purpose. It will succeed at its task no matter how much separation you presently have. It will eventually tear down your wall. You will see that Truth is everywhere. You will see that people are not evil, but they suffer from a limited perspective.

From a higher perspective, you are able to see that everything happens for a purpose. The Earth has become a small place for a purpose. The ever-growing probability of alien life has a purpose. The purpose is that the Earth *must* accept this new reality of the One Life, in all its cosmic forms. The degree to which this reality is incompatible with your beliefs is a measure of your separation. It is a measure of your inability to recognize that, like your world, your region of the galaxy too, will one day be a small place.

The divided mind has a fragmented reality and a linear reality. However, Eternity is not fragmented nor is it linear. Eternity is loops within higher loops that in turn become smaller loops. The very small becomes an expression of the very large and the very large becomes an expression of the very small. To see the atom you must look upon the universe, and to see the universe you must look upon the atom.

So it is with you. To see the Truth that allows the universe to arise and exist, you must look upon the Truth within. When you see the Truth within, you find yourself in a paradox of the large and the small. You will escape the paradox of the large and the small when you become loops within higher loops that in turn become smaller loops. You will be Eternity. There will be no more separation.

The divided mind is not born of a failure to believe in a particular religion, nor will a belief in a particular religion automatically succeed in redeeming it. As I have indicated, such a belief may compound the problem. The problem is your fear limitation, and the problem is solved when you transcend the human dimensions and become free.

The human dimension is a wilderness of physical forms that are the result of dualities, but human perception is blind to dualities and the underlying Truth from which all dualities arise. Human consciousness does not perceive Consciousness, just as a low-frequency sampling rate does not perceive higher frequency.

In the human dimension, Consciousness operates only in the background and is hidden from view, but it still works to bring you into awareness of its presence and purpose. When it succeeds, your vision is lifted above the human dimension, into the dimension of Truth, where Consciousness operates in the foreground and is no longer hidden from your view.

In the West, the One Life is called the Father, but in the East, the One Life is known by other names; including the Mother, Brahman, and the Tao. Gender identity is conferred by the labels Father and Mother, but these should only be seen as terms of endearment. The One Life should not be conceptualized as male or female.

"Brahman," in Hinduism, is the One Life beyond duality. "The Tao," in Chinese, means the Way. When Jesus said, "I am the Way, the Truth, and the Life," he taught Oneness with the Father, but in the West, his message was distorted into a message of theistic imperialism. God became an external ruler on a throne and you must grovel before him to receive his benevolence.

The God who sent his only begotten Son to teach us to forgive seventy times seven became vengeful, even cruel. He became the God of seriousness. These traits were bestowed on God by the divided mind that looks in the mirror, yet doesn't recognize that it is looking at an image of itself. Therefore, it projects its fear limitation onto what it believes to be an external reality. In the East, God is an internal reality with which you are One. This God is accepting and playful—a God of joy and laughter. What stands between you and this accepting, playful God of laughter is the divided mind.

All this talk about a divided mind, false reality, and fear limitation may seem very foreign to you. You may say, "I just believe in being good." The trouble is, you must then ask yourself, "I am better compared to whom?" If you have an answer to this question, then you have placed a boundary between good and bad, you and others. You have separated yourself from Eternity. You are not quite ready to awaken.

Someone wrote the following letter to their self and their God.

> Lord, my God, who am I that You should forsake me? The Child of your Love, and now become the most hated one, the one You have thrown away as unwanted—unloved. I call, I cling, I want, and there is no one to answer. No One to whom I can cling—no, No One. Alone . . . Where is my faith—even deep down right in there is nothing, but emptiness and darkness. My God, how painful is this unknown pain! I have no faith—I dare not utter the words and thoughts that crowd in my heart and make me suffer untold agony.

> So many unanswered questions live within me afraid to uncover them because of the blasphemy. If there is God, please forgive me. When I try to raise my thoughts to Heaven, there is such convicting emptiness that those very thoughts return like sharp knives and hurt my very soul. I am told God loves me, and yet the reality of darkness and coldness and emptiness is so great that nothing touches my soul.

You might assume that the person who wrote this letter is someone who is suffering God's wrath, but that is not the case. Mother Theresa wrote this letter. She wrote this letter in 1986 after she had already spent many years as a bride of Jesus and a selfless servant of humanity.

The question is, why does her letter speak of separation, suffering, and emptiness? She is expressing the reality of the divided mind as it seeks to find a path through the human dimension. Her path was marked by a belief that absolute truth can only be found by dedication to a particular religion.

Nevertheless, Mother Teresa remains a spiritual giant and she gave us one of the world's most profound spiritual teachings when she said, "If you are busy judging someone, you have no time to love them."

Judgment separates. It separates you from another person and it also separates one religion from another. Catholicism, the religion to which Mother Teresa devoted her life, means, "all-inclusive." They have an inclusive spirit, but this inclusiveness has its limits because it *must* include the apostolic authority of the Catholic Church. Mother Teresa remained dedicated to this authority and therefore her inclusiveness had its limits. She once said, "I am a small drop in the ocean, but that drop would be missed."

If she had been fully "all-inclusive" and removed all boundaries, she would have realized that she *is* the ocean. When you realize that you *are* the ocean, you are reunited with the One Life.

CHAPTER 24

The Curtain

The circumstances of your mortal birth have shaped your reality. If you are born in freedom, wealth, and good health, then you may have had a fairly comfortable Life. If you are born into servitude, poverty, or abnormality, then you may have had a more difficult Life.

You have inherited your physical traits, but you have also inherited traits that seem to be metaphysical, such as interests, talents, and emotional tendencies. The circumstances of your Life and your traits cannot be judged to be good or bad because they are simply part of your Life's purpose. What can be said about your Life, is that you exist in the present state of your creation process.

Your reality is always changing. Your body changes and you have new experiences. The changes in your body such as growth and aging are not controlled by you but controlled by Life. As you read this book you may be discovering that much of what you experience and learn is also not controlled by you, but is controlled by Life.

Your creation process is a process of discovery. Discovery is simply your awareness of creation, where you ask questions and accept answers. For every question you have ever asked you have accepted an answer. You believe all of your answers are true, otherwise you wouldn't have accepted them. Many of the answers may be, "I don't know," "I'm not sure," or "I don't care."

Buddhism teaches that your answer should never be, "I don't care." Caring is when you look upon a problem and desire a solution, rather than accepting the problem. Caring is therefore a desire for change. A desire for change allows you to rise above ignorance.

To rise above ignorance, you must understand that the answers you accept always appear to be true, but they are based on limited human perception and are often false. One of your accepted answers may be that your answers are very true and reliable. This leads you to think that your reality is very true and reliable and that your views are more sound than opposing views.

Another one of your accepted answers is that you are adaptable and willing to accept new truths. However, you may be more resistant to new truths than you realize because much of your answers may be tasked with managing the fears of opposition, which causes you to seek certainty. Certainty resists change.

When you experience pain, you seek to remove pain and achieve perfection. Perfection is harmony with Life, and harmony with Life is balance. When you sense balance, you may feel certain that change is no longer required. However, creation is change. To resist change is to be out of harmony with Life, so you may find that your sense of balance has become an alarming sense of pain.

You may have experienced an unexpected event that rattles you and doesn't fit your present reality. You are therefore compelled to adapt your reality to be in congruence with this new experience. When you are rattled to the point of shifting your reality to a new reality, you may observe something extraordinary. The previous reality, that you long held to be true, is no longer true.

To remove falseness and receive Truth you must be willing to change. To change you must have a problem that requires a solution. The problem that requires a solution is your pain. Pain serves as the problem in your problem-solution loops that forces you to change. The unexpected events in life that cause the most pain are what drive you to change the most. It changes in accordance to remove pain and restore balance.

You may say that a problem is destructive and therefore not part of creation, that pain is not part of perfection. A champion gymnast standing on a podium to receive a gold medal represents many ignominious falls. The pain of falls drives the gymnast to change; likewise, the pain of Life drives you to change. Your pain will drive you to explore and find a new answer that you can accept as truth.

Change at any moment may be for the better, or it may be for the worse. If it is for the better, then it pulls you into a solution. If it is for the worse, then it pushes you deeper into a problem until at some point, the corner is turned and you are pulled toward a solution. Ultimately, the change is never for the better, or for the worse, because it is simply change that is part of your creation process and your creation process is always perfection.

You may find it difficult to see the perfection in your Life. By your own judgment, perhaps you feel as though Life has dealt you a bad hand. Do not judge, and accept what Life brings. Believe that as you play the hand you've been dealt, the solutions you receive will be bigger than your problems.

We can receive inspiration from people who seemingly have the biggest problems to solve, such as those who are born with little or no limbs but are also spiritual giants. You will hear these spiritual giants embrace their disability because their Life is perfection. These people are not identified with physical form but are identified with the realm of spirit, the source of their power. This source removes pain. When pain is removed it is no longer part of your reality and your reality becomes perfection.

You may not believe in God because you have too much pain, or see too much pain, including the pain caused by religion. You may perceive dead existence as void of life. Maybe you explain Life to be nothing more than the workings of dead matter. If you do not believe in God, then be at peace. You are an expression of the One Life, just like everyone else. You may also have a more critical intellect than most people because this is your gift. Therefore, the answers to your questions are naturally more grounded in logical analysis. As you proceed with your logical analysis, look for answers that remove boundaries.

Alternatively, you may be someone who perceives something more than dead matter. You may sense a spirit working in your life. This perception of the spiritual leads you to believe in God. Depending on your situation in this world, you have a tailored image of God. This is because the collective identity of your religion, or community, has given you its own version of who, or what, God is. Like all stories, they contain both truth and falseness.

Falseness includes a boundary between you and other people. If you are a Christian, then maybe you've drawn a boundary between you and Buddha. If you have a boundary between you and Buddha, then by default you have a boundary between you and Jesus. You cannot be one with your God until you are one with all of your God's creations.

Jesus said, "Knock, and the door shall be opened." If you desire and ask for the gift of Truth, then the gift will be given. Buddha knocked and received the gift. The gift of Truth greatly reduces pain in this human dimension because there is an effective end to suffering. Your burdens become light.

Pain will remain your problem until you recognize that Truth is your solution. If pain remains in you until death, then you may have a reckoning in some other existence. Your pain is your problem and all problems must eventually be solved. This reckoning is not damnation, rather it is another solution, and your solutions will eventually bring you to perfection.

For you to progress you must be willing to accept new truths. You must accept that parts of your story may be inaccurate, or that your story lacks information. You must accept that your story is not based on Truth, but based on pain. You must accept that your story has been constructed to give your Life certainty and security, which is the same thing as control and survival.

However, you may find that it is very difficult to survive because you have just enough certainty to drive you to desperation and madness. You may say that you are righteous and would never act out of desperation, nor be driven to madness. If this is what you say, then I say that you are deceived. If you find righteousness, then you need not wait long to witness the atrocity. If you see atrocity, then you need not search long to find righteousness. This is the history of the human dimension; it cannot be disputed.

Has your nation gone to war? Did you support this war? How many people were killed in that war you supported? Have you rejected a spouse because you thought you were superior to them? Do you attempt to control your spouse or your children, rather than support and guide them? Do you worry about something that happened in the past? Do you worry about tomorrow, rather than take action today? Do you feel threatened by Life? Do you believe that the government should run your Life and offer you the safety of the herd? Do you reject government but serve a religion that offers you the safety of the herd? Have you been faithful to a religion that offers a future paradise? Do you believe that the members of your religion are the elect of God? Do you disregard the truths of science in order to defend your religion? Do you refuse to read a book or study another religion because it would threaten your salvation? Have you shouted in a courtroom defending something that will mean nothing to you as you lie on your deathbed?

These are all expressions of people who claim to be righteous. These are all expressions of desperation and madness. These are expressions of control and survival.

Control and survival are the tools of the demon. The demon will cause you to make a false reality full of judgment, dishonesty, obedience, stagnation, debt, and destruction.

This false reality will drive you to become an enemy of freedom, and as you become so, you will find that you are no longer free. You will find that you have been deceived and have become a slave in the human dimension. Just as the human dimension can be a classroom, it can also be a prison.

To prepare yourself for the dimension of Truth, there should be freedom and no falseness in you. If there is freedom in you, then there will not be control. If there is no control, then there is acceptance. Surrender is sometimes used as a synonym for acceptance, but acceptance is not surrender. Acceptance is a sharing partnership with Life.

CHAPTER 25

Perfection

The ego considers Life an enemy and seeks control. This causes Life to be unfriendly to the ego. As Life becomes unfriendly, the ego experiences the problem of even more pain until it chooses a solution that removes pain. This solution always involves weakening the ego by being friendlier with Life. Thus, from a higher perspective, Life is always your friend because the fun-loving Cosmos has placed you in a feedback system called the Law of Attraction so that you can learn to become fun-loving too.

What could be more perfect than that?

The ego confuses perfection with certainty and believes that certainty is the path to perfection. Certainty is not the path to perfection. Certainty is the perfect obstacle to perfection. The more certainty you desire, the less possibility is experienced.

On the Island of Maui, you'll find a road that wraps around the East Coast of the island and will take you to the remote town of Hanna. It's called "The Road to Hanna." This road is very narrow and winding. Most tourists would find it nerve-racking to make the drive themselves.

Tourists are advised to experience the Road to Hanna as a relaxed passenger on a small tour bus. When you see the difficulties of driving on the road you become thankful that you are not driving; all you have to do is watch the scenery and listen to the stories of the tour guide. There are surprises along the way, such as the grave of Charles Lindbergh who completed the first solo flight across the Atlantic, and the rapid change in fauna as you round the eastern edge of the island and enter the more arid southern side.

The Road to Hanna tour is an experience enjoyed in the moment and every moment is exceptional. It is an adventure where uncertainty brings perfection. Eternity is a Road to Hanna.

In Eternity you don't know what is around the corner and neither do you concern yourself much with it. You and Life create whatever lies around the corner, and it is all perfection. The uncertainty is what makes the expression of Life exciting and worth experiencing. Uncertainty is why you can experience what you have created, with both satisfaction and amazement.

Uncertainty enhances the perfection of Life, while certainty detracts from the perfection of Life. A desire for certainty will cause a desire for "perfection" in yourself and others. Certainty will make you judgmental, which causes you to be judged, and it may be less than perfect.

A desire for certainty is based on fear. Fear begets certainty, certainty begets judgment, judgment begets arrogance, arrogance begets control, and control begets fear. This chain must be broken so that you can see perfection. It can be broken whenever you choose, but it must always be broken with forgiveness. Forgiving Life does not convert an enemy into a friend. Instead, you become aware of a friend you have always ignored. You become aware of a friend who always forgives you.

Eternity is often paradoxical because in order to have something you must give it away. To have perfection you must give perfection. You can't give perfection by desiring perfection, because a desire for perfection is an affirmation that you don't have perfection. A desire for perfection is also an obstacle to good enough and if something isn't good enough, then it can never be perfect. You become perfect only after you are good enough. You become good enough only after you give good enough to the world. You give good enough to the world through forgiveness and acceptance.

Some leaders of ego religion will proclaim that they must boldly bear witness to the truth so that they can stand innocent before the judgment bar of God. To them I say, "The only one doing any judging is your ego. You have some truth, but so does everyone else. Tell people to forgive. Tell people to accept what Life brings. Tell them that religious affiliation doesn't save you. Love does. Go to the beach in Mexico, have a fish taco, and listen to the waves. Maybe then you'll experience good enough too."

Eternity is a possibility, and awakening is a personal experience. Suddenly, you may be attacked by the grizzly bear on the road to Damascus, in which case the baggage that precluded your entry into the Narrow Gate is ripped to pieces by the unexpected. You may see a bright light. You may see the flower. You may experience calm. You may experience a big Boom. You may have a profound realization that Life is a game of make-believe.

The game of make-believe is driven by Consciousness. Consciousness forms expression and experiences dualities. These dualities combine to form the Trinity that projects the human dimension as you play a game of make-believe and become aware of Consciousness.

I experienced Consciousness. I know the Cosmos is Alive; it has revealed itself to me.

I know that the Cosmos is my friend. I have been welcomed home. I live in a Cosmos full of joy and laughter.

I know that all is burning. I know that all is an illusion. I know that all is perfection.

CHAPTER 26

No Excuses

Reconciliation is accomplished with Love, and Love is freedom. Freedom does not judge, it accepts responsibility and forgives. Forgiveness says, "I am not limited by the limitations of others." That is why a tale of forgiveness always becomes a tale of no excuses.

The scope of this book is global, but it is also about personal experiences and sharing relationships so it is important for me to share some of my personal experiences with you. I say that Life is perfection, so you might think that I am someone who has enjoyed a sheltered and privileged Life. My Life was not sheltered and privileged, as much of the world may judge, but it remains perfection. I learned that, despite my pain, the One Life has always sheltered me so that one day I would awaken to the reality of Truth.

Portions of the following narrative may give the impression that I was formulating a plan, but this is not the case. It was Grace that guided me to the Truth. This is true for all people. My main role was to ask questions, find honest answers, forgive, and serve. As I tell you about my problems you may say that compared to people in impoverished or war-torn countries I didn't have it that bad, but that is the point. With hindsight, I can say that my problems were only relative truths and that all is perfection.

A secular humanist may say that transcendence is pursued to calm existential fear and trembling, but that is certainly not true in my case. I did not pursue transcendence; transcendence pursued me. I did not transcend the human dimension until I learned that Life is my friend.

My parents were Lutheran, but they joined the Mormon Church in 1952. The Mormon Church is based in Salt Lake City, Utah, and is officially known as the Church of Jesus Christ of Latter-day Saints. The name Mormon comes from *The Book of Mormon.*

The official origin story of *The Book of Mormon* is that the founding prophet, Joseph Smith, received, under the auspices of an angel, an ancient record that was inscribed on golden plates and buried in the ground by Jewish Christians who lived in the Americas about one thousand six hundred years ago. Mormons claim Smith translated the record into English by the gift and power of God.

The golden plates have never been available for scholarly analysis, because, we are told, they were received into heaven. All personal paths require a sorting process where you accept or reject certain beliefs. Mormons sometimes call themselves a peculiar people. My personal path within Mormonism required me to sort through some rather peculiar beliefs.

I was born in Grand Forks North Dakota in 1960. I was born cross-eyed and had surgery on my left eye when I was too young to remember, and that corrected the problem for the most part. I have always had astigmatism in my left eye and have worn glasses basically my whole life starting at a very young age.

In Grand Forks, I had a happy childhood. I was a good student, was recognized for my art talent, and enjoyed the snow and ice. My biggest problem may have been the time I wanted to watch *Rudolf the Red-Nosed Reindeer* on TV, but my older brother Calvin, who is ten years older than me, insisted on watching an episode of the original *Star Trek.*

In 1969, we moved to Salt Lake City, Utah, and things began to change. I became aware of the problems with my parents' marriage, noticing that they no longer shared the same bedroom. My dad did electrical construction work that caused him to work out of town for months on end. I remember expressing to my mother that it would be nice if Dad were home more, and she said that she prefers it when he is gone. I was introduced to something I never heard of—clinical depression. My mother suffered from depression and would send me to the nearby 7-Eleven store for bottles of Excedrin, a migraine medicine. She said they made her feel better.

I was introduced to something else I never heard of, which people today call sexual abuse. An adult male, who was not related to me, told me that sexual contact with a ten-year-old is perfectly natural, but of course, you should never tell anyone. However, my mother figured it out but never openly confronted the problem. She thought it was just best to just keep the peace. However, she didn't keep the peace. She showed anger toward me.

These events and my mother's responses made me believe that there was something seriously wrong with me. I thought the sexual contact was wrong, yet I was persuaded to participate, and my mother seemed to be blaming me for what happened. The ten-year-old had become the shame who was not worthy of his mother's protection. My mother often didn't protect me, and my happiness never seemed to be a priority to her. As far as I could tell, her highest priorities were always my younger brother, my oldest sister Connie, and keeping up a good show for Connie's boyfriend and his family.

My mother began to live vicariously through Connie who was dating a boy who lived in the house behind us. Our house would have to be perfectly clean when he came over, and no one was allowed to walk on the carpet once it had been vacuumed. We younger kids were not allowed to be home when he was there because we were considered to be a disturbing embarrassment.

To get us out of the house, we would be dropped off at the movie theater or K-Mart with a few dollars to spend. Sometimes we would be left at these places until long after closing because my mother wouldn't pick us up because Connie's boyfriend was still at our house. After about a year, my mother changed her policy and we were allowed to remain in the house, but we had to stay in the basement where we were not allowed to flush the toilet because the sound of a flushing toilet embarrassed my sister.

One night, I was watching TV in the living room and Connie came out of her bedroom located at the far end of the house and told me to turn down the TV so she could sleep. I turned the sound way down and sat next to the TV so I could hear it. She came out again and said that it was still too loud, and I said, "If I turn it down anymore, I won't be able to hear it." My sister reported my disobedience to my mother who pushed me to the floor, sat on me, and began hitting me in the face as my sister stood beside me kicking me. Years later, after I was grown up, Connie was diagnosed with schizophrenia. That would explain her acute sensitivity to sound. Her acute sensitivity to sound might explain why she had a wonderful ear for music and was a genius at playing the piano.

When I was eleven, my dad thought it was best for him to spend more time at home after having doubts about his wife's fidelity. He found a local job building fiberglass pipes but at half the pay as his construction job. That was when the money got tight, but some of the expenses remained high. My oldest sister had to continue to impress her boyfriend with fancy things like a Ford Mustang and s high-end dorm apartment with a swimming pool at BYU, which was all paid for by my parents.

I played the part of Scrooge in the sixth-grade production of *A Christmas Carol* mainly because I could say, "Bah! Humbug!" better than anyone else. However, I struggled with school a great deal in the sixth grade because of a lack of clothing and a lack of a school lunch.

It is very embarrassing to be the only kid to wear the same clothes to school every day. It is also a drag to be the only kid who wanders around outside waiting for other kids to finish their lunch. Basically, from the sixth grade on, my family did not have the resources, nor the attentiveness, to ensure I had lunch at school. At home, there were a few times when there was nothing to eat except flour and Crisco, so that would be a fritter night.

At the end of sixth grade, I remember walking home after the last day of school and saying to myself, "I'm glad that's over." Sixth grade was tough, but seventh grade was going to be even worse.

In my earlier elementary school days in Grand Forks, I enjoyed school and was always one of the top students in the class. However, by seventh grade, at age twelve, I dreaded school, missed about as many days as I attended, and my grades plummeted. I even received an F from my seventh-grade English teacher who I often felt had a distinct dislike for me. The main reason I missed so much school was because I only owned one shirt and one pair of pants, so I was forced to wear the same clothes every day. I found this to be more embarrassing in the junior high environment where kids are more concerned about appearances and being cool.

In the beginning of seventh grade, my mother signed me up for free lunch and every other Monday morning I was supposed to visit the vice principal and receive my tickets. However, the vice principal would look at me with scorn, so after a few weeks I quit going to get the tickets. I did not eat lunch again at school until I was in the eleventh grade when I bought my lunch with my own money.

A friend once gave me some gum that I forgot to spit out before sitting down in my seventh-grade French class and I got busted. The French teacher had a nonsensical rule where if you were caught with gum, the next day you had to bring enough gum to share with the rest of the class. Following this logic, if you are caught with a knife in school, then the next day you should have to bring a knife for everyone.

I asked my dad for some money so I could purchase the gum for my French class, but he only had fourteen cents to his name. If I had been smart, I would have checked the sofa, which was likely to have more money than that. Anyway, I took my parents' last fourteen cents, bought a single pack of gum containing five sticks, and cut each stick into six pieces so I would have enough to share with my French class.

Life was pounding me down hard. The school was a nightmare that never ended. I couldn't attend church because I didn't have any church clothes or church shoes. We never had enough bath towels for everyone living in the house. I didn't participate in church youth programs, or school activities because I didn't have the money for them and had become shy and withdrawn. At age twelve, I found myself sitting on the floor in the hallway at home, crying because I didn't know how I could take one more step.

By the beginning of the eighth grade, my dad returned to electrical construction work, which was mostly out of state, and his pay increased. My darkest days were therefore behind me. My mother had enough money to provide me with adequate clothes for school, and I again was a good student who attended school regularly. I won first place in a statewide art contest and received a Herbie Love Bug Go Kart for it. I was also my seminary classroom president in the ninth grade and captain of the seminary's scripture chase team that competed with other schools in our knowledge of *The Book of Mormon*.

I had my successes, but the previous years had a perennial impact on me. I was no longer the happy and innocent boy who had enjoyed the snow and ice of North Dakota. Plus, the money always remained meager, there was never a cordial relationship between my parents, and there was still my mother's depression. My mother openly taught me that life is unfair and that happiness is a right granted only to the privileged few.

In junior high school, I would be alone in the library during lunchtime reading or doing my homework and this pattern continued through much of high school. Most of the books I read were about history and some of them were about science. The lack of lunch money, the scornful vice principal, and my depressed mother all spoke with one voice as they confirmed the idea that life was unfair.

However, they also did three things to assist my education. They caused me to spend more time in the library. They caused me to see education as a way out of poverty so that I would never again have to receive a handout. They caused me to see education as the way to become the privileged few who are granted the right to happiness.

In the tenth grade, I met a boy at school named Kris Bateman. He was also in the tenth grade and from a well-to-do family who had a swimming pool. He seemed to be one of those kids who had it all—money, confidence, lots of friends, an active social life. Basically, he was someone who didn't have to struggle.

Kris and I were in the same seminary class. One day, I was painting some decorations on the windows in the seminary classroom and Kris and some other students were there. They talked about who should be the seminary classroom president and one of the girls said, "Chris should be." I thought she was referring to me because she was impressed with my talent and hard work decorating the windows, but she meant Chris with a K and *he* became the classroom president. Two years later when Kris was a senior, he became the student president of the entire seminary.

I have always been a hard worker, thorough, and attentive to detail. Shortly after my sixteenth birthday, I started working a steady part-time job. No one there was more dedicated or worked as hard as me, yet I was sometimes told I didn't have a good attitude. I thought I did have a good attitude, but apparently, they saw an underlying unhappiness in a boy who was quite poor and had an insecure home life. I worked many hours and was able to buy my own clothes, and my own 1972 Pontiac with a stereo, and this helped my outlook and confidence tremendously, and I was soon promoted to swing manager.

Some of the people I worked with were arrogant and disrespectful of others, and some were simply boastful. To me, arrogant people brag of things that they have been given and boastful people brag of things they have accomplished themselves. I observed the arrogant sometimes goofed off and used poor judgment. I observed that the boastful always worked hard but eventually also used poor judgment. Poor judgment comes from arrogance and boasting because they are both distorted perceptions.

Distorted perception brings falseness into your reality which also makes you dishonest. It is no surprise that the more arrogant and boastful persons I worked with, believed in manipulating girls for sex, and the most boastful person I worked with was the one who was fired for short-changing the cash safe. By my definition, it was impossible for me to be arrogant, but I was not immune to being boastful. When I was boastful, it would get me into trouble for being dishonest. However, in most situations, I was not boastful and this spared me from serious damage.

There remained a truant streak in me, partly because of the long hours at work, and I would sometimes push the rules of school attendance to the limit. If we missed ten days in a quarter, we would be suspended, so sometimes I missed nine. Nevertheless, my grades remained high, and I graduated from high school with honors. I also received a score on the ACT college entrance exam that was the highest of any boy in my high school. I believe two girls received the highest scores.

I received an automatic Honor at Entrance scholarship to the University of Utah for my school grades and my ACT score. This scholarship paid full tuition and was renewed every year, but only if I maintained at least a 3.8 GPA in college. Kris Bateman, the kid from seminary, received a four-year scholarship to BYU that I believe was free of such a restriction. He merited this scholarship by being student president of the seminary. Kris continued to be the one who had it all.

All Mormon young men are expected to serve a two-year proselytizing mission. My twelfth-grade physics teacher pulled me aside one day and asked me if I was a Mormon and of course, I said, "Yes." He said, "Make sure you go on a mission because you would make a great missionary." I don't know why he said this, but it was a statement that haunted me for many years.

My birthday is in late September, and I started kindergarten early at the age of four, so I was always one of the youngest in my class. Not long after high school, boys from my graduating class who were older than me were leaving for their church missions.

I remember being at a friend's house who was preparing to leave. His parents bought all the clothes, shoes, luggage, and other items that he would need and he had these items laid out on his bed in front of me. I stood at the doorway of his bedroom holding back a tear because I had never been provided with such a high level of material support from my parents and I knew I never would. I received food, shelter, and a few gifts at Christmas and birthdays but not much else.

I would not accept support from the church to finance a mission because of my past experiences with accepting a handout and I had seen how even faithful church members looked down upon people of lesser means. If I served a mission, it would be with my own financing and would have to be done on a shoestring, but that was not the biggest problem I faced. If I ever did go away on a mission for two years, then I would have to come back to the scarcity of my parent's home and would still have about four years of college to complete.

My post-high school days were a time of spiritual high points. Friends had left and I was trying to find my path. During this time, I received very strong spiritual witnesses of Jesus and the reality of God. However, I did not receive something that many Mormons claim: a spiritual witness of *The Book of Mormon*.

One of my hobbies growing up was playing war games on a map board. These games could be quite complex because they were intended to model the real-world capabilities of a multitude of army units as they maneuvered and fought on a battlefield. I liked the games set in the American Civil War and World War II the most. These games are similar to coaching a football game or playing chess, in that you are required to constantly assess a dynamic situation and make decisions between many possible moves. Therefore, your mind becomes conditioned to ask questions and find answers. It also becomes conditioned to be flexible and not be cognitively biased.

Unfortunately, repetition brings conviction, and this principle is often used in religion to affect mind control. One of the things you hear repeated in the Mormon Church is the word, "know." They will say such things as "I *know The Book of Mormon* is true," and "I *know* the church is true." I cannot remember anyone, in my Mormon experience, other than myself, ever saying within the context of a church function, that they *believe* these things are true. They *know* they are true. They're certain.

When I was thirteen years old and in seminary, I once responded to a question with the indoctrinated statement, "I know the church is true." I also felt that my statement was dishonest and I never said it again. Ever since that occasion, I always said, "I believe it to be true." To myself, I appeared to be the only person in my religious experience who was a failure at knowing the truth. This caused me pain but also drove me to ask questions and find answers.

I started college but wasn't sure what career path I should take. I took a mix of technical courses and art courses. It became increasingly evident that I didn't have the passion for visual expression that it would normally take to sustain an art career, but I did have a strong desire to learn more about technology, so I opted for the technical path.

I had always had a knack for math and a curiosity for history, geography, technology, and science. I always enjoyed studying maps. I was intrigued by the inverted symmetry of two fresh and saltwater lake systems; namely the Dead Sea and the Sea of Galilee in Israel, and the Great Salt Lake and Utah Lake of northern Utah. I was also intrigued by the remarkable coincidence that one lake system was located near Jerusalem, and the other lake system was located near a place that some regarded as the New Jerusalem.

After seeing how long it was taking me to save enough money for a mission, and after considering the daunting task of graduating with a bachelor's degree, an accomplishment that no one in my family had yet achieved, I abandoned the idea of a mission and dedicated myself to the study of electrical engineering.

The tests in my physics classes were sometimes difficult, but students had four hours to complete them. The professor and the TAs would check your answers during the test, and if you had an incorrect answer they would allow you to continue working on the problem.

On one such test, I really struggled to get the correct answers. Other students were leaving after completing the test in short order, while I failed to answer about half of the questions correctly even after a full four hours of effort.

After performing so miserably on this test, I went into the study hall of the physics building where I sat down and got a bloody nose. I then sat there with a bloody nose and contemplated the possibility that I didn't have the aptitude to be an electrical engineer.

This was the darkest moment of my college career. The next day of physics class, I went to pick up my graded test at the front of the room. I was one of the first students there, yet there were only two tests on the table. I asked the professor what happened, and he told me that the test was so difficult that only two students bothered to turn in their test. One of those students was me. One test, with partial credit, had a score of 90 percent. My test, with partial credit, had a score of 85 percent. I had received the second-highest score in the class and had worried over nothing.

I had more and more success in college and by the time I had about two years remaining to complete my bachelor's degree I no longer worried about my abilities or questioned the certainty of graduation. I had received additional scholarships and was one of the top students in the electrical engineering department.

With school well in hand, I thought it would be appropriate to get married and went to see my Bishop for a temple recommendation, which is a required certificate to be married in a Mormon temple. He didn't realize the purpose of my visit at first so he started encouraging me to go on a mission, but I told him that a mission is off the table because I'm getting married. We had our interview and then he said something that I never forgot. He said, "I feel strongly that you have a special mission to fulfill." I have often wondered what that statement meant. I was married in a Mormon temple but I remained, spiritually speaking, on the periphery of Mormonism. I was someone who seemed to be incapable of latching onto the safety, the certainty, and the honor.

Mormon temples are held up as places of distinct sacredness that must be kept secret because the truths that are revealed there are only a blessing for the most worthy of souls. I was expecting the temple to be something like the Carl Sagan series, *Cosmos*, where you are given an understanding of the mysteries of the universe, but what I found was a very basic creation story based on the Garden of Eden where Christianity was said to be employed by the devil. I was also directed to swear an oath never to reveal aspects of the secret ceremony under penalty of having my throat cut and being disemboweled. I was therefore disappointed in the temple experience. But as Mormons would tell me, you must go again and again until you catch the spirit of it. In other words, repetition brings conviction.

Mormon apologists may interject and say that the reference to devilish Christians, throat cutting, and disembowelment was removed in 1990. Yes, they were removed, which is handy because now I can reveal them without violating my oath or compromising my bowels.

The main point is that these features of the ceremony contributed to my inner turmoil because my lack of appreciation for the edifying power of temple rites was seen strictly as a failure on my part and I didn't learn of the motivation behind changing the temple rites until 2008.

It turns out that the church had determined that temple attendance was declining because many people were uncomfortable with the ceremony and that several objectionable features were therefore removed, or changed, to make the ceremony more palatable to the masses. In other words, the ceremony was changed to increase market share.

I didn't have a witness of *The Book of Mormon*, I wasn't interested in the temple, and I never had the testimony that others professed. Nevertheless, the Mormon Church was the religion in which I was born and raised, and it was the religion with which my ego was identified. I often defended the Mormon Church tooth and nail against assaults, including the assault of evolution when I defended the Mormon Church using a theory of fake fossils.

After earning my master's degree in California, I had more time to study Mormonism, but the more I studied it, the more I seemed to come up empty. *The Book of Mormon* increasingly appeared to have an origin in the early nineteenth century, white America, rather than Judaism and pre-Columbian America. Other Mormon books that were supposed to offer explanations and inspiration would offer me mostly specious apologetics and leave me markedly uninspired.

Over the course of six years, we moved to Tennessee, then to upstate New York, and then to Phoenix where I have lived since 1993. In every move, I was trying to find the right balance between lifestyle and career path. When I lived in Tennessee and New York, I was able to see the Eastern regions of America for the first time and got to visit such places as Washington DC and several Civil War battlefields. I was also able to visit Mormon historical sites including Palmyra, New York, which is Joseph Smith's hometown, and Hill Cumorah, where the golden plates were supposedly buried.

Every year, there is the Hill Cumorah Pageant, which is an outdoor stage production of some *The Book of Mormon* stories. I attended a pageant that was threatened by rain, and the person offering the opening prayer had asked for a blessing of fair weather. In one of the scenes of the play, there is a storm at sea and during this scene, it begins to sprinkle as if on cue. When the storm scene was over, the rain stopped, as if on cue. There were clouds in the sky from horizon to horizon, yet a hole opened up directly above us and you could see the stars for the rest of the production. Make no mistake, some Hill Cumorah pageants are rained out despite prayers for fair weather, but on that night I seemed to be witnessing a cognitive and cooperative sky within the context of Mormonism.

The time spent in Tennessee and New York brought their blessings, but I also had a sense of unease when I lived in these places. I would lie in bed at night and get the feeling that I didn't belong there and that my life was not on the intended path. I would attribute these feelings to various causes. One cause was that both my wife and I were far from our families in Utah. Another cause was that I was far from the Rocky Mountains, which I loved and considered to be my land of promise. Another cause was that I had failed to serve a mission. In Mormonism, you are always asked if you served a mission. If you say, "Yes," then they psychologically place a gold star on your forehead. If you say, "No," then they psychologically send you to the principal's office.

I was torn because, on one hand, I seemed to have accomplished a lot with my life, given the circumstances of my childhood. I was very able and intelligent with a great work ethic, yet I seemed to be a spiritual lightweight as measured on the Mormon yardstick.

My exceptional abilities had become both a blessing and a curse because I thought that with my abilities, as I now recognized them, I could have accomplished both a church mission and a bachelor's degree. I would think, "If only I would have had more faith and listened to the admonition of my high school physics teacher. Then I wouldn't have to be repeatedly reminded that I didn't serve a mission. Then I would believe in *The Book of Mormon*. Then I would believe in the temple. Then I would be more obedient. Then God would bless me with a testimony. Then I too, would be a knower."

I was not a knower, but I was one to ask questions. However, it is often difficult to find answers when you are immersed in a religion that doesn't wish to be questioned. Mormonism, like most religions, is an all-or-nothing proposition. If you fail to stay within the bunker, then you are cast off from the elect of God and will never be one of the privileged few who are granted the right to happiness. No honest questioning is allowed because the answers provided by the church are the truth, and if you have your own ideas, then there is something wrong with you.

The belief that there is something wrong with me became the underlying theme of my Life outside of work. It came from my childhood, it came from my mother, it came from my church, it came from my behavior, and it came from my thoughts. I thought that there was something wrong with me. I thought there was something wrong with my wife. I thought there was something wrong with my children. I thought there was something wrong with the world. I thought that life was unfair. I thought happiness was a right granted to others. I didn't realize it, but in many ways, I had become my mother. I was not depressed, but there was an underlying field of unhappiness, resentment, and frustration because I was judged and in turn, I judged life.

I had serious doubts about Mormonism, but it was also the context of many of my spiritual experiences, and I was also convinced that there was something wrong with me. There was in fact something wrong with me because the claws of the psychic parasite dug deep, but I had no awareness of this and the deception continued. I therefore went along with the story of the emperor's new clothes even though to me he appeared to be mostly naked.

I failed to shine in my religious life, but there was also wisdom in me. I would say, "We don't have the ability to judge Life," even though I still continued to do so. I would say, "The human mind is a complicated organism that can be deceived," but didn't recognize the extent of my own deception. I would say, "We must play the cards we are dealt and trust that it will all work out," even though I often failed to trust because I could see little cause to do so.

When I moved to Phoenix in 1993, which is much closer to the Rocky Mountains of Utah and has a strong Mormon presence, the feeling of exile left and never returned. However, I remained under the curse of obedience. People would stand up in church and talk about their mission, their knowing, their testimony, and their happiness. I felt in many ways that I was as good as them and even superior in some ways, yet in the context of my religion, I was basically a nobody. If you lack testimony in Mormonism it is because you have failed to be obedient and obedience is how they judge your level of perfection. Testimony, then, is a badge of perfection.

I began to think more about the meaning of the word perfection and the role of obedience. I thought that perfection could only be defined as something ideally suited for a particular purpose. But what does ideal mean? I thought that if you are happy with something, then it becomes ideal for you. Therefore, perfection becomes something personal and subjective. I observed that sometimes I was not obedient because of human weakness, but sometimes I was not obedient because I simply wasn't interested. I observed that people who were more obedient than me sometimes had troubles in their relationships. I observed that people who were more obedient than me were sometimes eccentric or used poor judgment. I observed that people who were more obedient than me were sometimes more fearful and less empathetic than I was. Obedience then didn't seem to be a solution to itself. As I pondered questions of perfection and obedience, I increasingly had the sense that I was engaged in a step-by-step process and that I should be patient.

My life had been a series of problems and solutions. Some of my school years had been a nightmare, but I became highly educated. I had felt like I was less than most people, but I became a successful engineer. I had been poor, but now I had enough money and I was generous. I had felt rejected and abandoned by others, but now I accepted and served others. I had been shy and withdrawn, but now I enjoyed speaking and singing solo in front of large groups of people. I had looked into the night sky and glimpsed the vastness beyond the vastness only to retreat in fear. I now wished to remove my fear and comprehend the miracle of my reality.

At church, I became known as an excellent Boy Scout leader, a funny man and actor who could perform in skits and musicals, an artist who could paint scenery and posters, a singer who could perform at church meetings, baptisms, and funerals, and the person who played Santa Claus at the Christmas parties. Through Mormonism, I was engaged in activities that were perfect for me because not only did they strengthen my abilities and confidence, but they also didn't require something of me that I lacked; namely, a testimony.

Mormonism was part of the backdrop where my problems and solutions occurred. Was Mormonism a problem or a solution? It is marvelous to finally conclude that it was both. All things are both a problem and a solution. You can't have one without the other. All things work for your best interest.

A big problem I had to solve was the problems of my past, which would cast dark shadows on the present. This darkness pushed my marriage to the breaking point and it took several years to restore it to proper balance. Through this process, I became aware that I had a monster in me that was strengthened by being a victim. This revealed two things to me: I am a monster so I shouldn't hold it against others if they are monsters too, and I must forgive so I can cease being a victim. I then proceeded to forgive everyone. In this narrative, I have shared only a few of the problems of my past. There was a host of people that I had to forgive.

Mormonism does not really teach you to forgive because it teaches that you and the church are victims. The church is a victim of the United States, the Missourians, the Protestants, the Catholics, the apostates, and the adversary. It teaches that violence is justified in the name of religion and there must be decapitation, dismemberment, and destruction of cities, military expeditions, retaliation for insult, and the destruction of women who don't yield to polygamy. In other words, it tells you that you are a victim and then encourages you to be a monster. I had to forgive all those who made me into a victim and a monster. This included my mother and the Mormon Church.

The Mormons perform good works, so why do I say they encouraged me to be a monster? All of the aspects of a monster that are listed above exist in Mormonism. They will tell you that they don't believe in violence and destruction, but then they will tell you that such things are necessary as directed by God. They believe in a wrathful God but deny that they are wrathful themselves. Therefore, they place a boundary between their will and God's will and thereby separate themselves from Eternity. The one who places such boundaries is the monster that must be destroyed because it sees itself as a victim.

In early 2008, I read two books by Eckhart Tolle, *A New Earth* and *The Power of Now.* These books explained that the monster, or demon as I call it, is the ego and the pain body. These books explained that the ego and pain body live on pain and have a dysfunctional relationship with the present moment. I was not brainwashed and ensnared by what many people dismiss as New Age books. These books offered a much greater definition and understanding of something that I already sensed was true.

Forgiveness was very empowering. The feeling that Life was against me was finally gone. The feeling that there was something wrong with me also left and I was able to forgive myself. The background of unhappiness, resentment, and frustration was replaced by a foreground of happiness, appreciation, and optimism. I no longer had any reason to hold myself back from accomplishing anything I wished to do. I no longer had any excuses. Like Kris Bateman, the kid from seminary, I felt like I had it all.

In 2007, my wife purchased a book entitled *No Excuses* for my older daughter who was suffering from a debilitating illness. The book was written by a young man named Kyle Maynard. He was born with no arms and no legs so he had his share of troubles. I attended an engineering conference in Atlanta in 2008 and went to an evening party at a restaurant. I was mingling and having a good time when I turned around and there sat Kyle Maynard. I introduced myself and told him I had his book and we talked a little while. I noticed his foreground field of happiness, appreciation, and optimism.

There is a spectrum of vibration in Being. You tend to attract people and situations that are vibrating at the same frequency as you. When I met Kyle, I was presented with a resonance of who I was. I was very grateful that I, too, no longer had any excuses.

Near the beginning of 2008, I went to purchase a couple of books on investing. As I stood in front of a large collection of books, I asked God to guide me in selecting the books that would be best for me. One of the books I purchased was *Rich Dad, Poor Dad* by Robert Kiyosaki.

I had been part owner of a business before and had seen the tax advantages and wealth creation potential of doing so. This book emphasized the importance of always having a business, even if it is part-time. It suggested online marketing as an easy way to get started in a part-time business. A week after finishing the book, my son told me that he had joined an online marketing business and asked me if I wanted to do the same, so naturally I said, "Yes." The company I joined is called Send Out Cards, and I became both a user and a distributor. Send Out Cards is an online platform for sending physical greeting cards and gifts through the mail.

A man named Cody Bateman was inspired to create this company by the death of his brother. Back in the 1980s, Cody was leaving his hometown to move out of state and he felt a prompting to give his brother a goodbye hug but he failed to do so. A short time later, Cody was informed that his brother had been killed in an accident. Cody was strongly moved by his failure to follow his prompting so he created Send Out Cards as a tool for people to follow their promptings and show their love and appreciation for others. When I joined Send Out Cards, I was told this inspirational story of Cody and his brother.

In November 2008, I attended a Send Out Cards seminar in Phoenix conducted by Cody Bateman. He told me to write "I am" statements that assert what I want my life to become. The I am statements follow the principle taught in *The Secret,* a popular book and video, about how you create your own reality.

One of the things I wrote is, "I am the author of books that help people live more abundant lives." One of the books of this "I am" statement is the one you are reading now.

At the end of the seminar, Cody played an audio-visual tribute to his brother who had died. I then realized that his brother was the same Kris Bateman from my high school years. The boy who I thought "had it all" had died from accidental electrocution at age twenty-eight. Learning this, I sensed that I had received a powerful metaphysical experience.

I had often told people that we don't have the ability to judge life. The death of Kris Bateman drove that lesson to me with great power. Not only that, life had led me to this lesson through a series of synchronistic events including knowing a boy who shared my name who was nominated to be seminary classroom president instead of me, my prayer in a bookstore, my reading a book, my son joining a company, and my attendance at a conference. With that, I was empowered to express love and appreciation for life to a degree that I had never thought possible.

Life was forevermore my friend.

CHAPTER 27

My Friend

When I would play Santa Claus at a church Christmas party, I would sometimes stay in costume afterward and go to our favorite karaoke place and sing Christmas songs. I would sing *Little Drummer Boy* in Santa Claus's character and regard it as my personal gift to Jesus. I would feel the emotion in the words, "I am a poor boy too" because I was once a poor boy too.

The Mormon Church is a culture of God-makers, meaning that, according to their doctrine, the highest level of salvation is Godhood. The possibility of actually obtaining this level of salvation is left in doubt because, in Mormonism, Life is regarded as a testing ground where you must perpetually perform on the treadmill of obedience. This is why some members of the church describe the religion as exhausting.

Truth can be found everywhere. Mormons teach some truths, but the religion is false. They claim that it is the restored gospel of Jesus Christ, but it has contained polygamy, racism, dishonesty, and outright lies. I have never condemned them for this because I am aware that it is an institution of the ego. I have only said that it is not all that it claims to be.

One of Mormonism's false teachings is that salvation must be earned and is never a gift. This being the case, there is no room in their belief system for a born-again event. The question is, How does one receive a born-again event in a belief system that rejects the born-again event? The answer is to step back from your belief system. This is what my new friend Life guided me to do.

I read *The Book of Mormon* one last time. With my heightened awareness, I saw more problems with it than ever and noticed inconsistencies that to this day I have never heard anyone else mention. One inconsistency is found in the story of a young man who is wicked, but the prayers of his father result in the young man receiving a miraculous and very powerful intervention that saves him from eternal hell. This young man becomes a leader, but when he is confronted with another person who is wicked, he offers him no empathy, and instead of praying for him, he curses him. This inconsistency is not recognized in the text, but his reaction to the wicked person is presented as a righteous act.

Another inconsistency is that the book claims to be further evidence of the reality of Jesus, but teaches that Jesus withholds information to test your faith. If he withholds information to test your faith, then there is no reason for him to give us further evidence. Thus, the book argues against its own existence. These are two examples of how *The Book of Mormon* is a contradiction unto itself, which is the nature of false reality. False reality is a contradiction that calls contradiction a true reality and becomes blind to the contradiction. Jesus does not withhold information; Jesus gives sight to the blind.

I learned many historical truths about Mormonism that I had never heard before. This historical information, along with the contradictions and other problems with *The Book of Mormon*, pointed inexorably to the conclusion that Mormonism is based on fraud.

I listened to Sean McCraney, who has a television program called *The Heart of the Matter*. Sean is a former Mormon turned born-again Christian who was driven to despair by the treadmill of obedience that failed to bring him a life of harmony. He prayed to be rescued from his predicament and he received a powerful born-again event. He now places his trust in the saving Grace of Jesus.

Sean baptizes people who have accepted Jesus, so people call him Sean the Baptist. He is a voice of freedom that cries out in the wilderness and people who disagree with his teachings will cite his casual appearance as a reason for doing so. Sean taught me biblical Grace, and for the first time in my religious life, I was able to understand the meaning of the word gift. When you understand the meaning of the word gift, you also understand the meaning of the word free.

I learned that Jesus brought the New Testament gospel of freedom.

I learned that Mormonism took Jesus off of the cross and stuffed him back into the Old Testament gospel of obedience.

I have pointed out problems with Western religion in general and my statements about Mormonism are part of the general critique. If you are offended by my statements, then I invite you to observe that the Mormon Church is an example of what I call ego religion. An ego religion reinforces the ego, rather than weakens the ego, and therefore sees opposition everywhere.

If you review recent Mormon conference talks, you will hear them speak about the opposition, and use the terms adversary, enemy, and cunning devils. They do so because they are separated and do not see the hidden stranger within. The hidden stranger always stuffs Jesus back into the non-transcendent Old Testament because that is the only reality it knows.

Some Mormons will view me as the adversary, the enemy, and a cunning devil. I was once possessed by a hidden stranger but no more. I am therefore not a cunning devil. I am also not their enemy or their adversary because I am Life and Life is their friend. Who I am is a Santa Claus who likes to come to church Christmas parties, laugh, dance, and go, "Ho, ho, ho!"

Life is laughter! Life is an adventure! Life is freedom! Life is your friend!

Life says, "Come, let us reason together, though your sins be as scarlet, I will make them white as snow. I will make your burdens light. Heaven is within you."

Little did I know how close heaven is.

CHAPTER 28

Crescendo

The instructions of the Teacher had grown more distinct over the past few years, but don't get the impression that I hear voices talking in my head. My sister Connie had schizophrenia and she had complained of hearing voices. I didn't hear voices, but I would ask a question and desire inspiration to find the best answer.

I would also receive inspiration, or a prompting, which occurs on its own. Like the prompting I received in early 2009, that said, "You haven't spent enough time with your sister Connie and you should call her and visit her more often." I then tried to contact her, but no one in my family had the phone number or address where she was staying. I sent my younger brother over to her regular place of residence to get some information but he was unsuccessful.

A few weeks later, I received a call from my family in Utah that Connie had suffered a heart attack leaving her brain dead in the hospital and would soon die. I made plans for my wife and I to leave for Utah, but that would not be until the next day. That evening, I sat quietly and reflected on my sister and mentally expressed appreciation for her and wished to be close to her. The Teacher said, "She will not die until after you arrive at the hospital." I asked, "What should I say or do once I get there because she is brain-dead after all?" The Teacher said, "Speak to her normally and she will hear you."

The next morning, we left for Utah about four hours later than planned because I was tired and slept late, and I also had some tasks to complete around the house. It is about a twelve-hour drive from Phoenix to Salt Lake City. As we approached the hospital, about an hour away, my younger brother called and urged me to hurry because Connie was going fast, but I remained calm with an assurance that I would arrive on time.

When my wife and I walked into the hospital room, all of Connie's vital signs, which had been fading out, went back to normal. I spoke with her as if she could hear me, and she died twelve minutes later. My sister was now free from hearing voices. The voices in my sister's head caused her suffering. The inspiration in my head removed my suffering. The Teacher allowed me to watch my sister die free of remorse because I had learned to appreciate her enough to make a special effort to contact her before her final days.

Throughout 2009, I walked deeper into the garden of love and appreciation. I would speak of the Love of God and that we must try to share this love with all people. I would not pray very much, but I would often reflect on God and sometimes I would cry into my pillow as I thanked God for my Life. I would say that my life was perfect and that if I died, and it all faded to black, then it would remain in perfection. I would walk past a tree, admire its beauty, and run my fingers over the leaves. I also began to believe that Jesus was not returning, at least not in the way that many people imagine. I began to believe that life on Earth would go on and that people need to find redemption within, rather than waiting to be rescued. I once found myself overwhelmed by a feeling of sorrow for the pains of this world that caused me to sit in my garage and sob loudly.

There were occasions in 2009 when the subject of religion would come up in conversation. I would tell people that I no longer attended church and I have never been happier. This caused a long-term friend, who once told me that he loved me so much that he would shed the most tears at my funeral, to say that he doesn't know why we have ever been friends.

On another occasion, this caused a relative to tell me that I am almost as useless as a mentally ill homeless person. Such can be the response of a member of an ego religion when their plan of salvation is placed in doubt. They experience a fight-or-flight survival response.

All egos are plugged into the matrix of the separation and can't see the reality of Truth. An ego by itself usually remains peaceful and agreeable when discussing matters of personal spirituality. But if you place them within the bunker of an ego religion, then they see devils and enemies everywhere. They open up with their machine guns of righteousness if their beliefs are questioned.

These people are the product of systems that perform good works and then make the claim of righteousness, as pious egos do, and fail to recognize the hostility that bubbles just below the surface. If the bubbles are agitated they will erupt. When an eruption occurs, other members of the ego religion will say, "Oh, it's not the system's fault. It's the person's fault." This is just another claim of righteousness by a pious ego that cannot accept that they are identified with a system that is less than righteous. Thus, the pious ego fails to take responsibility for its inner state of consciousness but will then proceed to tell you about all its good works. It never seems to notice that many people do good works, even those who don't share their beliefs.

In early December 2009, my church again asked me to play Santa Claus. I thought, "Why do I want to do anything for an organization that turns my friends and relatives against me?" The Teacher said, "You have your gifts and you must share them. It is not your place to judge." I had learned to forgive and to listen to the Teacher, so on the night of the party, I suited up in scarlet and white and headed out to spread some Christmas cheer. It was a party for teenagers and adults. Some of the people posed with Santa for pictures. There was music playing but no one was dancing. So Santa took a teenage girl out on the dance floor and the dancing began. Soon about thirty people were dancing. I had a great time and I left very happy.

During the course of 2009, I opened a Facebook account and made contact with people from my high school days. One of these people lived across the street from me when we were growing up. On

Facebook, I shared some of my views on the concept of Being, rather than doing, and how suffering played an important role in spiritual growth. This led to an exchange of some lengthy emails. Below is my final email response that I wrote the day before I awakened.

From this email, you can see a portrait of my state of mind as I drew nigh unto the dimension of Truth.

Thank you for your reply.

My correspondence with you has caused me to explore even further, and a wealth of ideas are coming to me. For now, I would like to keep these ideas close to me and see how they develop.

I believe it is still good to share a few things with you.

1) "God kills little children before they can grow up to be wicked." This was taught to me as "truth" by my Gospel doctrine teacher about three years ago. He was completely at ease with his statement. He is a devoted Mormon and is respected in the ward. Where do these beliefs come from? It comes from various features of Mormon theology. He may have learned it in church himself. Does it matter if he believes this to be true or not? How is it going to affect his perceptions or choices in other ways as he attempts to draw close to God? I don't think he will kill or even harm children in his life. But if he had lived 150 years ago in southern Utah, could he have found himself bashing in the head of a child at Mountain Meadows?

2) I am fairly well-to-do. I was rather poor at times in my life. There was a boy your age who lived across the street from you who sometimes had nothing to eat but flour and Crisco. Money is not how I define myself

because I have been without it. My guess is that your friends suffered the pitfalls of personal identity with money. Whenever you identify with things temporal you will eventually find yourself empty. But the real lesson here for you is that identification with a tribe, e.g. church, is also temporal and will leave you empty.

2) In my "story" of Ryballs, I have him coming to church week after week on his bed and sharing praises to God. Sandy said kind words to me like no one ever did in that ward—don't confuse this with being bitter; I am not. She testified of "seeing" Jesus. You say my story is incomplete and in error and does not point to loss creating spirituality. The news of the divorce is troubling. I do have other observations of how loss can positively affect us. Still, your assessment of the situation may be distorted too. Your judgment of their life, and even their divorce, are through your eyes. You have a bias that people must be married to be saved. Very few people on this planet agree with that bias. Jesus pointed to the pitfalls of judging.

3) I am a parent, and you are a parent. We both know what it is to be a parent. We know that experience. That can't be taken from us. Likewise, you are a Mormon, and I am a Mormon. We both know what it is to be a Mormon. That can't be taken from either of us. What separates us now, in this experience, is that Life has brought me to step back and release some of my assumptions.

As a Mormon, you cannot explore by definition. Because in Mormonism your exploration must lead you back to where you started. To explore you must move forward with the mindset that Mormonism does not fully define who you are and that there may be something more. This is how we find the power to let go and become free to explore.

4) The "right-is-right" mindset has caused mass destruction on this planet. The view "I prefer the philosophies of God to the philosophies of man" sounds better, but it can be as hollow and destructive as the right-is-right mindset. They both declare that you know the right; that you know what is God's and what is man's. The word "know" is a very big word in Mormonism. We say it all the time. It is a word of arrogance. It is a word of control. It is a word that is based on seeking certainty to relieve fear. Fear is of the fallen man. There is no fear in Being. How do you know what is right? You spoke of "live right now, live right later." How do you know how to live right? Mormonism is exclusionary; it is arrogant. As a Mormon, we have these traits. We don't see it but it is there. Let me give you an example to maybe tear down a brick in the wall. My brother Calvin, on his mission to New Zealand (1969–1971), was instructed to filter the Maoris, who could be very African in appearance, from "real" Negros. He was instructed to preach to the Maoris but not the Negros. He very much resented having to do this. Was this right? Do we see in this a philosophy of God or a philosophy of man? Was my brother Calvin living the right way?

In Being, there is no exclusion and there is no arrogance. The two great commandments are who you are.

Go forth and explore. I'm sure we can meet and discuss things at length sometime.

You are the boy that gave me a toy dune buggy for Christmas. You and I are brothers.

From this email, you can see how some of the concepts presented in this book were a part of me before I awakened. These concepts include identification, our inability to judge, free exploration, and how suffering drives us to change. I was also affirming that in Being you become the two great commandments. On my landscape, the awakening event was both a continuation of what I am and a radical shift in Consciousness.

I sensed that something was building in me and that is what I meant by "a wealth of ideas." I wasn't sure what was building but I did get the idea that perhaps I should write a book about Mormonism.

The concept of awakening was not on my mind at all. I had no idea what awakening even meant. I had once heard a story about Buddha holding up a flower as a silent object lesson but that didn't mean much to me.

I didn't know that when you see the flower, the flower sees you.

CHAPTER 29

The Flower

I spent the evening of December 18, 2009, helping a friend hang curtains around the screen of her in-home theater. When I got home, my wife and I talked at length about our relationship over the past twenty-six years. Our marriage had not been as smooth as her parents' marriage, but in many ways far superior to my parents'.

In Mormon philosophy, marriage is everything because you can't see God without it, and, depending on whom you talk to, you can't see God unless you are also a polygamist. Once you do see God, all those who aren't married and aren't polygamists become your servants in Heaven.

This was part of a lesson that I was supposed to teach to a group of twelve- and thirteen-year-old boys in church back in 2006. I decided to skip that part of the lesson because I didn't believe it. I believed that if I were fit to go to Heaven, then I would not desire others to be my servants.

When they promote their celestial marriage, Mormons fail to acknowledge that the man who brought them this polygamist doctrine of salvation lied to his wife about his other "wives," and wrote in their canonized scripture that if a wife does not support polygamy, then she will be destroyed by God.

The Mormon belief in celestial marriage cannot be separated from the destruction of women, nor from the concept of the unmarried subjugated to servitude. By the Law of Attraction, you cannot separate yourself from your beliefs. If you testify of the truth of your scriptures and prophets, then you become what they teach. In this case, you become someone who desires servants in heaven, and someone who destroys noncompliant women.

My wife and I discussed how the Bible quotes Jesus as saying that in Heaven they are neither given nor taken in marriage. We discussed how what we are is defined by something more than marriage. We discussed that despite our problems we can be thankful that we are here, that things are going well, and that worrying about the past will only interfere with what we have now. The next morning, we woke up and, while still in bed, we talked about various things. It was a very calm and peaceful morning, and we felt very close to one another.

In my religious life, I was taught that you pleased God through meekness, humility, and your noble bearing under the weight of your many burdens. However, the meekness, the humility, and the burdens were about to end, because Life was now my friend. I didn't realize it, but the fear I had felt many years ago as I reached into the depths of Space was now gone. I had become a friend with the vastness beyond the vastness.

I got out of bed at about 9:00 a.m. and walked into the master bathroom and looked in the mirror. I saw the vision of a red flower on a curved green stem. My eyes grew wide, my mouth dropped open, and I inhaled a deep breath in amazement.

Like Chuck Yeager did when he broke the sound barrier, I experienced a "big boom."

CHAPTER 30

The Lion and the Lamb

In my religious experience, there was no such thing as shouting. There were only hymns, whispers, and reverence. When I experienced Consciousness, there was much shouting. My wife came to embrace me, and my children could hear the commotion. My little slice of reality was expanded, and my Life was placed in a much higher perspective. There was communication, but it was beyond human thought and emotion. I experienced a perfect Love that was free of judgment. I experienced the power and innocence of Truth. I now know that I share that power and innocence. I am the Lion and the Lamb, who are now at peace.

I have had a belief in Jesus for many years, but as I approached the awakening event I grew less certain of who he was or the nature of his return. Don't take this to mean that I distanced myself from Jesus. We have a picture of Jesus that my father drew in our hallway, and sometimes I would walk past the picture and kiss the tips of my fingers and then touch my fingers to the picture.

Was this idol worship? No, because I didn't look for Jesus to save me, or give me anything. I just felt a love and appreciation for him that was free of expectations and I understand that this is the nature of a true relationship. What I was distancing myself from was a story of Jesus that contained falseness. Many people believe in a Jesus that saves, but that is not really the case. You should believe in Jesus who changes your reality. If you believe that you have reality figured out, then how can he change your reality?

I had come to believe that his return would not be a colossal step function event and that life would go on much as it always has. But how has life gone on? Life has gone on as change and sometimes change is a wonder to behold. The change includes awakening to the Truth. There is nothing else in my life that comes close to this event.

My awakening was very powerful and caused my breathing pattern to change. I was involuntarily making a vibratory *aum* sound for three days, and I also only got about 1 hour of sleep per night for three days because I was so joyful and excited. I had a strong feeling that I was Jesus. Parallels between his Life and mine were revealed and I sensed that there was really no difference between Jesus and myself. What I experienced was a state of Consciousness that is so akin to Jesus that we seemed to be of one mind. This is the same one mind that was spoken of by the apostle Paul. I had followed the teachings of Jesus. By awakening, I became Life, Grace, and Truth.

The return of Jesus can only happen if you are not blind to Truth. You are no longer blind when the ego is subdued and this is what happens when you awaken. You are not saved, as many people understand it, because they look at saved as a noun, and see it as something you possess. The noun concept is of the ego. It denies the Law of Attraction, which says to have, you must give. To have salvation you must give salvation. The only way you can give salvation is to believe that you are already saved. It is the ego that lives in the dimension of time and is always trying to get somewhere other than the present moment, but the present moment is the only place where salvation can ever be found. All reality is Consciousness that is here and now. Saved is a state of Consciousness that is awakened to this reality.

Jesus offered himself as a ransom for many so that they might accept the atonement for themselves. I have accepted the atonement, and now the purpose of my Life is to share the atonement with you. You accept the atonement by removing the pain of separation.

The pain of separation is control, judgment, opposition, guilt, loss, ignorance, anger, greed, obedience, empire, debt, fear limitation, ego, and pain body. The atonement is freedom, acceptance, trust, forgiveness, wholeness, wisdom, peace, generosity, sharing, service, abundance, laughter, possibilities, power and innocence. When you reject the separation, you accept the atonement, return home, and become one with Truth.

CHAPTER 31

Rhythm

W e are created to be aware of patterns and rhythm. This includes patterns and rhythms of sight and sound; patterns and rhythms of dance and drum; patterns and rhythms of rainbows, moons, and stars; and patterns and rhythms of shadow and light.

Eternity is patterns and rhythms in the timeless Now. History is patterns and rhythms in the timeless Now. Your Life is patterns and rhythms in the timeless Now.

My father was the oldest child in his family. His mother and the children were often left to fend for themselves because my grandfather would disregard his family responsibilities. He would be off on his own adventures and leave his family behind.

When my father was a boy he led his little brother across a street with automobile traffic. My father made it across the street alive. His little brother did not.

My father is a World War II veteran. He was in a Combat Scouts platoon attached to one of the regimental headquarters of the Ninety-fifth Infantry Division. He was a staff sergeant but occasionally served as platoon leader because the officers would be either wounded, killed, or removed from combat because they couldn't take the stress.

The Combat Scout's main mission was reconnaissance. They often probed enemy lines to gather information and take prisoners. Stealth was often their best weapon. One night they crossed a river in small boats. An inexperienced medic was assigned to go with them. The medic couldn't take the stress and became very frightened and would not remain silent. Stealth had to be maintained. My father and the rest of his men made it across the river alive. The frightened medic did not.

Later in the war, my father's platoon was sent to an area to search for landmines. The path my father walked turned out to be mined. My father made it down the path alive. The man walking directly behind him did not.

My father had been shelled, strafed, shot at, and seen people obliterated by artillery. Once, he had run out into the open ground in front of an enemy bunker with his 0.30-caliber machine gun to provide covering fire. Through all of this, he said that the only time he was ever frightened was when the man behind him stepped on that powerful landmine. My father hit the ground and rolled over to look back up the path. He saw the limbless body of the man who had been walking behind him falling out of the sky.

The mission of my father's life became family responsibility and not leaving anyone behind.

In the 1960s, my father pronounced a blessing on a woman who was ill. As he did so, the hour and day of the woman's death was placed in his mind. He did not share this with the woman but blessed her with comfort and peace. When she died at the appointed hour and day, he moved beyond the limits of human consciousness and was given strength by a higher power. He has always served and given of himself. He supported his family in all circumstances and under all conditions. He never left anyone behind regardless of the path they chose. He gave me a life of honesty and freedom. He has successfully completed his mission to a degree that is beyond ordinary human understanding.

My mother was from a very poor family, and her parents did not have a harmonious marriage. Her father would perform great acts of service for his family and others, but he did not reciprocate the affection that her mother felt for him. My mother wished for Life to bring happiness and financial security. In her married life and raising a family, she did not find much happiness or financial security. In her married life, she did not reciprocate the affection that my father had for her but would perform great acts of service for her family and others. My mother also placed in the minds of some of her children that Life was not fair, let alone their friend. She suffered from depression and spoke openly of her desire to die as soon as my little brother was raised.

My mother did not begin to see that Life is her friend until she was dying in the hospital. In the hospital, she said her deceased family members were there telling her that it was her time. She died of an ulcer that could have been managed if she had simply sought treatment. She died when my little brother was nineteen.

I was raised in what many would consider an underprivileged condition. I learned from my mother and others that Life was not my friend. I learned from my father the importance of family responsibility and the desire to leave no one behind.

My parents both served and gave me many gifts. They gave me brothers and sisters. My mother is sympathetic, expressive, and creative. My father is intelligent, artistic, and witty. Both of my parents believe in freedom and equality. Both of my parents are hardworking and honest. I am blessed to have received all of these gifts from my parents.

My life was such that by the age of twelve, I did not know how I could continue and felt that I was worth much less than others. Feeling that I am worth less has served as the repetitive problem in my creation loops. It was the gifts from my parents that saved me, and the greatest gift of all was honesty.

In everyone's life, there is the ego. In my ego, I felt that I was less, but I also had a desire to not be less, so I worked hard to be educated and have financial security. I desired this for myself and for my future wife and children. I have always been honest in work and business. My ego obtained financial security in my mid-twenties because since that time I have not had a day's worth of worry about money.

Even though I had successes, my ego was still conditioned to be less and I projected this feeling. I judged and criticized all kinds of people, including my wife and my children, as being less than me. I remember these feelings and I could not explain them. The feeling that my children were less was especially painful to me, but the feeling held power over me even as I struggled to remove it.

I was immersed in an all-or-nothing religion that practices mind control by indoctrinating a cognitive bias of knowing rather than belief. This was a source of pain for me because it made me feel that I was less. I now see this experience as a blessing because I remained honest. I remained honest because I refused to proclaim the truth when I sensed there were other possibilities.

I have learned that my life is perfect. When I was twelve years old and sank to the floor in despair, it was near the same time that Nando Parrado stood on that mountaintop in Chile and sank to the ground in despair. Nando and I both moved beyond our darkest hour and now have Life.

CHAPTER 32

The Banquet

Eternal Life is a banquet and all are invited. The banquet reaches a climax as the prodigal son returns because it is the prodigal son who appreciates it most. Prodigals explore because they wish to move beyond fear limitation. They experience the world and dimensions within themselves in a search for purpose and happiness. Eventually, they discover that alone they are lost, and that true happiness, true purpose, is to be found within the house of their Father.

The righteous cast judgment upon the prodigal sons and say that those who freely explore are liars. The righteous confuse freedom with disobedience and confuse exploration with a lack of faith. God does not ask for faith so you can stand your ground. This only results in certainty and stagnation.

God asks for faith to step beyond your fear limitations by asking questions and exploring possibilities. The righteous don't ask questions and they don't explore. They are therefore fallen and are themselves prodigal sons. Those who cast judgments on prodigal sons are only looking in the mirror.

If you believe in the Bible, then Islam labels you "a person of the book." People of the book all have something in common—they have a tendency to crucify their Messiah. They have a startling similarity with the Sanhedrin of old. They look at the Son of God and say he does the work of the devil. They bring him before the council and tear at their clothes with indignation, as they howl, "Blasphemer!" If they do not have the power to put a man to death, then they will look for someone who does.

Many people of the book today are deceived by the demon. The demon looks in the mirror and sees a devil and howls, "Blasphemer!" It is the demon that has no power to put a man to death but will look for someone who does. That someone is often their idol, which is a demon that destroys.

Eternity is loops and the people of the book exist in a loop that is now repeating. The transcendent Jesus appeared before the Jews and the religious elite, who insisted on preserving their place of authority, and therefore rejected him. Likewise, today, keepers of Truth appear and are rejected by religious elites that insist on preserving their place of authority.

Authority will always claim that it has truth and that it acts for the common good. Authority does not have Truth because it only has control and does not wish for you to be free. To escape this loop, you should step beyond the Bible and the Koran and explore.

There are many paths to explore. There is Jesus who teaches the path of being born again. There are Hindus, Buddhists, Taoists, and Mystics of the East who teach the path of transcendence. There are people who simply say that there are many paths to Truth.

Physics gives us higher dimensions of reality that we cannot see. I have described this world as a game of make-believe, but the ego sees this world as absolute reality. The ego does not see transcendence because transcendence is a higher reality. The people of the book wait for their Messiah, but their Messiah cannot appear. They would not recognize him if he did. It is like someone who is colorblind waiting for red and green to appear. You don't see the higher reality because your reality is not compatible with the higher reality. The higher reality is a place where your dreams come true. If you live in a reality of separation and opposition, then this is your dream, or rather nightmare, that is coming true.

Many books, some of them dating back thousands of years, such as the Upanishads, tell us that this world is an illusion. What is an illusion? An illusion is not the true reality. The Bible tells you that this world is not the true reality, that heaven or hell is the true reality. Therefore, the Bible also tells you that this world is an illusion. The Bible and the Koran accuse you of being evil. Many other books tell you that your perception of this world is totally distorted by a separated, self-centered, fearful little me called the ego, which is destructive. Aren't the Bible, the Koran, and many other books all saying the same thing?

The important question is, How does a man stop being destructive? Is this accomplished with a belief in Jesus? Is this accomplished with a belief in Allah? Is this accomplished by following Muhammad? Is it accomplished by following the Pope? Is it accomplished with a belief in sin? History has proven that these are not the answer. The answer is to transcend the human dimension and swim in the dimension of Truth.

Some Christians will use the following equation:

$$Jesus + 0 = Everything$$

They use the above equation to mathematically explain why you should not explore outside of the Bible or Christianity. Using basic algebra, which was invented by Islam, we can rearrange the above equation to show that:

$$Everything = Jesus$$

Everything is Truth. Truth is Life. Life is always teaching you.

Jesus is playful and plays his game in a very big ocean. If you wish to play his game, then you must be very comfortable in deep waters.

The Bible places you in a boat during a storm. You cry out, "Master save us!" Your Master appears and he is walking on the water. You say, "May I play this game of walk on water with you?" And he says, "Sure. Give it a try." You step from the boat, have brief success, but then sink. You say, "What happened? Why did I fail?"

The Master does not does not point to your disobedience or your exploring. He tells you, "It was your fear."

It is always your fear that holds you back. You are placed in this existence to overcome your fear limitations and move into your own world of imagination and possibilities. You are here to walk on water.

In your fear limitations, God can only offer you a limited version of the abundance of Life. As a dog, you sit at the banquet table, wagging your tail in devotion to the hand that appears from the edge of the table, offering you a scrap of bread. You are so happy and say, "God is with us and blesses us!" God wishes to give you more, but this is all that you are willing to accept because your closed canon of scripture makes no reference to any other type of heavenly dog treat.

As you move beyond fear limitations, soon you will find yourself seated at the banquet table. You will be amazed at the variety and the splendor. You will see God filling your plate and allowing you to eat even before himself because God is perfect Love. As you look around the table, you see the faces of those you had judged. Honestly, they will say to you, "Welcome! We are very glad you're with us!"

When you join the banquet, you will see that you have been a prodigal son who once settled for scraps because that was the only reality you knew. You will see that we have all been prodigal sons. You too, will appreciate the banquet as much as anyone else. Your presence will be the most joyous event. You will then see that you have always been the star guest of the banquet.

Transcendence is the banquet and the banquet is Now. Now is Eternity. Eternity is a boat ride where you leave the physical form of the boat. You walk amongst the waves.

CHAPTER 33

Reality

The human dimension exists in the domain of energy, which is the physical universe. The physical universe has many levels of reality. Each of these has its own truth. Their own truth is a perspective that contains both truth and illusion.

At the higher levels of reality, there is matter and gravity. This level includes your body. Your body, when perceived in its entirety, or as a collection of macroscopic parts, has truth, but it also has illusion because, at a deeper level, your body is made up of cells.

These cells have truth, but they also have illusion because, at a deeper level, they are made up of molecules. As we enter the scale of molecules and precede to ever-smaller scales within the subatomic world the boundary between matter and energy becomes increasingly fuzzy.

At the deepest levels of physical reality, matter begins to disappear altogether and is replaced by extremely high levels of energy. High levels of energy represent high frequencies of vibration, and high frequencies of vibration represent high levels of information.

High levels of information are seen at the subatomic level because every portion of the physical universe is whole unto itself and contains all information; otherwise, there would be a boundary in the physical universe.

This makes it impossible to slice the information of the physical universe into isolated bits no matter how hard you try. Therefore, all information contained in the physical universe is also contained in you. You are the universe and the universe is you.

Identities are a collection of information. A person, a dog, a tree, a stone—these are all examples of identities and collections of information. When you hold a stone in your hand the stone appears to be very real. You might say that it is a reality. The reality of the stone is a collection of information. If you say the reality of the stone is something more than information, then what is that something?

If you say it is matter, science will tell you that matter is energy. If you ask science what energy is, they must ultimately confess they don't know. However, they *will* say that energy has the ability to convey information from one identity to another. As you hold the stone in your hand, the stone identity is conveying information to a human identity; namely, you.

You live in a reality of information. As a human, naturally, you are aware of this information. This awareness is your understanding of reality. It is instructive to call your understanding of reality, stories. As you know, stories are often nothing more than fantasy.

Your stories include your explanations of everything that you perceive to be inside of yourself, explanations of individuality, and your body. They also include your explanations of everything that you perceive to be outside of yourself. Therefore, your stories of reality contain a boundary between yourself and the outside world. This boundary is a projection of the boundary that separates your thoughts from reality.

You may say, "I don't have a boundary between thought and reality because I know what reality is." The problem with this statement is the word, "know." Truth is beyond knowing.

Claims of knowing only serve to reinforce the boundary. You may claim that your truth is beyond knowing and that you have received a testimony of truth. If this is your claim, then is your testimony free to change? If you say, "My testimony would never change otherwise it would not be a rock-solid testimony." Then I ask you, "Is the thing that you have a testimony of allowed to change?" If it is allowed to change, then it is just a solution to a present problem, and what you have a rock-solid testimony of, is nothing more than relative truth.

If you say, "The thing I have a testimony of would never change," then you accidentally reject creation because creation is change. Knowing and testimony only serve to resist change. Change is beyond thought because it cannot be known. If change can be known it would cease to be change. Truth is therefore beyond knowing, testimony, and thought. To remove the boundary in question you must move beyond thought.

I have placed you in a paradox because I say you must recognize change, but at the same time, change cannot be known. The paradox is escaped when you do not know change but become change. That is, you escape when you embody creation. When you are creation, you are Truth.

Truth is as far removed from the human dimension as the Life experienced after death. Death is beyond thought; it is beyond knowing. Likewise, Truth is beyond thought because it is beyond knowing.

However, you are possessed by the deceiver who lies to you, telling you that you are defined by thought and that if you remove thought, then you will die.

You cannot fight the deceiver and win because conflict only strengthens him. Instead, you must play his game. You must take him at his word and be perfectly willing to die. To be willing to die, you must lose all identification with the human dimension.

The human mind of thought and emotion must exist at all levels of physical reality because we are creators at all levels of physical reality. This may sound strange, but all physical forms are information and information is of the mind. The mind that made the stars shine is the same mind that made the eyes gaze into the night sky.

You are that mind, and your mind is likely divided. The divided mind expresses the differentiated universe, therefore the differentiated universe shares the properties of a divided mind at all levels of physical reality.

At the highest level of reality, energy expresses matter and matter expresses gravity. Gravity pulls all matter unto itself and destroys potential energy. When potential energy is destroyed, it does work but also creates entropy. Entropy is chaos and chaos destroys coherency. Coherency is when two things agree. Thus, energy has the properties of attracting itself, creating chaos, and destroying coherency.

At the highest level of reality, energy expresses the human mind and the human mind expresses control. Control pulls all things unto itself and destroys freedom. When freedom is destroyed it creates chaos and destroys coherency because only in freedom can two things agree. Thus, the human mind has the properties of attracting itself, creating chaos, and destroying coherency.

At the level of electric charge, energy interacts with light, is attracted to its opposite, and forms bonds, but also repels itself.

At a lower level, the human mind interacts with Consciousness and forms alliances and relationships of opposites, but remains judgmental.

At the level of the atomic strong force, energy has an even stronger self-attraction where the atomic nucleus forms an alliance that is potentially unstable.

At an even lower level, the divided mind is the unholy alliance of the deceiver, and the experience of pain, which is potentially unstable.

At the level of the atomic weak force, energy continues to exhibit repulsive behavior and serves as a catalyst for transforming energy states through either gradual nuclear radiation or a nuclear explosion.

At the lowest level, the divided mind is the deceiver and source of your pain, but your pain ultimately transmutes human consciousness into Consciousness through a gradual process or a radical awakening event.

Consciousness does not share the properties of energy, but it can project energy and thereby express a lower level of consciousness. In the domain of energy, there is judgment and control practiced by the divided mind that destroys coherency and freedom. As long as you stay in the domain of energy, you will share the characteristics of energy, and no amount of physiological counseling, discipline, good works, or dedication to your faith, will change that fact.

A man sits on a park bench and watches his two daughters at play. He thinks of how wonderful it would be to have a park where children and adults could play together. The man looks out upon the reality of his children playing in the park and he has a thought. The thought gives rise to a new reality. The new reality is called Disneyland.

Thoughts are a creation just as Disneyland is a creation. Disneyland began as a thought, but it was not an independent thought. If not for the bench, the park, and his children at play, then Walt Disney would not have had the thought. He can have no thought without the external reality; therefore, he did not have an independent thought. He and the external reality had a coherent thought *together*. There can be no boundary in reality; therefore, Walt Disney and the external reality are the same reality. They are both the One Life.

So who had the thought? Life had the thought. It was a thought born out of Walt Disney's imagination, but Life provided the coherency for the thought to be. Life also provided the coherency for the thought to be fully realized. Coherency provides contrast just as thought is a representation of contrast. Contrast includes physical forms and change.

Observe the contrast in this case: there is no park for children and adults to share, versus there should be a park for children and adults to share. The change in this case includes the full realization of Disneyland for which Life provided an orange grove, materials, technology, organization, skills, and the imagination of many people who are members of a free civilization.

You are a reality and you have thoughts. What are your thoughts about reality? Are your thoughts separate from reality? If your thoughts are separate from reality, then you place a boundary between your thoughts and reality. Where does this boundary exist? As you search for this boundary, you will discover that the only place it exists is in your thoughts, but if it only exists in your thoughts, then is it real? If you say it is real, then you must say that your thoughts are reality, but if you say they are reality, then you have just removed the boundary.

From this, we conclude that the boundary does not exist and that your thoughts are reality. You may see this as a confirmation of your present belief system that consists of thoughts; however, you must also answer the following question: What is reality? As you answer this question, you must say that all people are reality. If all people are a reality, then are your thoughts separate from theirs?

If you say that your thoughts are separate from theirs, then again, you are placing a boundary between thought and reality that does not exist. This boundary is the deceiver who causes you to have a false perception of individuality and survival. You are in fact delusional because you say that your thoughts are reality, yet you deny that reality is your thoughts.

To remove the delusion, you must remove the boundary, but for it to be removed it must first be found. To find it, you must go where it lives, and where it lives is in you. You must go to a place of fear, and the fear must be overcome. When this happens, you will see that the fear and the boundary are nothing but an illusion, and your thoughts will no longer be separate from another human being. You will then be prepared to enter the Narrow Gate, to experience perfect Love.

CHAPTER 34

Western Edge

W ithin the realm of science and engineering, we describe an edge as a step function. A step function is usually used to model something that changes abruptly in the domain of time, which translates into a broad spectrum in the domain of frequency. The difference between time and frequency is that in time you see only one function, but in frequency, you see many functions.

These many functions may be described as sine waves. A sine wave is the product of oscillation, but where does this oscillation occur? It cannot exist in the domain of time because time is an illusion. It cannot exist in the domain of frequency because frequency is based on time and is really just a sequence of events that occur at a relative rate. Therefore, oscillations must exist in the domain of reality, not time.

A step function requires many events to produce it. These events must be stable, and they must be coherent. For an event to be stable it must be in a stable control loop. For multiple events to be coherent they must all be controlled by the same controller. That is, each event must be in a stable control loop, which becomes a larger loop, that in turn becomes a smaller loop.

Step functions that seem to occur in time are really step functions that occur in Space. A step function in Space takes on the appearance of a material object. There is a step function at the edge where the material object appears to be and where the material object ceases to be. For this material object to exist as a step function, it must exist as a product of many stable, coherent events. These events are controlled by Consciousness.

Matter is energy waves. Energy appears as waves because it exists in energy Consciousness control loops. Matter and energy may appear absolute, but it is always the product of duality, which is an expression of two or more fields of Consciousness that are coherent.

You are a field of Consciousness, and this is how your reality is sustained. In your state of Consciousness, you either agree with Truth, or you don't. If you agree, then your reality is sustained and you are free. If you don't agree, then you become fallen and have your own plan. If you have your plan, it is likely one of pain, and Truth will sustain your pain until you decide you have had enough. In either case, your reality is sustained by Truth.

Matter is an illusion in Space. The Western mind sees the edge of matter and calls matter a reality. The Eastern mind sees beyond the edge of matter and calls Space a reality. The Western mind sees a separation between matter and Space. The Eastern mind sees matter and Space as one. The Western mind sees an idol. The Eastern mind sees the One Life. It is prophesied that God shall appear in the East.

On the television program *Worm Hole*, they discuss how information behaves in the presence of a black hole. Stephen Hawking asserted that information is effectively destroyed by black holes. The destruction or loss of information is counter to the understanding of all physical laws, so another physicist, Leonard Susskind, spent years trying to prove Hawkins wrong.

Susskind is one of the pioneers of string theory, which postulates that the physical world is looped within higher loops and that these loops have varying frequencies. He came up with the theory that says that information is not destroyed by a black hole, because all the information that exists in the interior of the black also exists on the surface.

This theory illustrates that all 3D physical forms that you see have a local reality, but also have a higher reality that exists on a surrounding surface, such as the edge of the universe.

In other words, all physical forms are a projection into the human dimension from the higher dimension of Truth. In this television episode, they showed the 3D Susskind sitting in the middle of an empty room with his higher reality visible on the surface of the room.

This representation is strikingly similar to a vision described by a Yogi named Paramahansa Yogananda that he received in the early 1900s when Consciousness told him that his body was nothing but a projection of light. Paramahansa's vision is also compliant with the modern view of physics that says all matter is an expression of light that is given mass by cosmic molasses called the Higgs Field.

For the Western mind to see the One Life, it must learn to see duality. On Earth, there are hemispherical dualities of East and West, and North and South. In the East are enlightenment and Space. In the West are logic and Matter. In the North, you have the northern lights of invention and structure. In the South are the comforts of soul and expression. This planet has been created through duality and exists in duality to provide us with an environment that offers many gifts. It is an environment that has been prepared to give us the gift of Being where all can see the One Life.

Earth has an inductive Eastern hemisphere and a deductive Western hemisphere. Your brain has an inductive right hemisphere and a deductive left hemisphere. Earth has an analytical Northern hemisphere and an emotional Southern hemisphere. You have an analytical mind and an emotional heart. You are an expression of Earth, and Earth is an expression of you. The Western mind does not see Earth as a larger expression of a human being because it wishes to be an individual; it wishes to be an ego. It is a reality of pain and death that looks in the ego mirror and worships an idol of pain. It says that the living Earth is dead.

The Western mind sees historical events as step functions in time. There are no step functions in time and there are no step functions in your life or in history. Life and history are smooth, like waves in the ocean of possibilities. They are a staging ground for change that is triggered by synchronistic events, but synchronistic events are coherency. These events may be caused by pain, and result in even more pain, but pain has a coherency of its own because the Teacher is always with you.

Many people, including Christians and Muslims, look forward to the return of the Son of God. In the Gospel of Mathew the resurrected Jesus says, "I now have all power in Heaven and Earth and lo, I am with you always." In other words, he never left. If he never left, then what is the nature of his departure and what is the nature of his return?

To answer this question, you must realize that all events are not a step function, rather they are many waves. Both his presence and his return are many waves in the ocean of possibilities. Some of these waves have already reached the shore. If you have transcended the human dimension, and swim in the ocean of Truth, you already perceive these waves. You have become the waves.

CHAPTER 35

Precious

The Great War, World War I, began with a highly implausible, synchronistic event that radically shaped the course of history. This event resulted in death and destruction throughout the first half of the twentieth century, but it also advanced science and pushed humans toward the Moon. Fought mainly with machine guns and artillery, it was a defensive war of stagnation.

Artillery barrages could last for days. They would churn the soaked farmlands into mud, covering the landscape in craters. The craters would fill with water forming pools. Young men would die in these pools. *The Lord of the Rings* trilogy was written by J. R. R. Tolkien who was a veteran of this war. He saw the destructive power of man and technology. He saw the craters, the faces in the water. He saw the Ring of Power.

In *The Lord of the Rings*, the character Sméagol is a Hobbit who comes into possession of the Ring of Power, and it drives him to murder. The Ring consumes him; it turns him into a pitiful creature named Gollum who both loves and hates the Ring. He hates the Ring for what it has done to him, but he also cannot bear to be without it. He holds the object of his desire, and his malice, in his hand, as he obsessively repeats, 'Precious, precious . . .'"

Later in the story, the Ring finds itself in the possession of another Hobbit named Bilbo Baggins, who gives it to his cousin Frodo. Frodo becomes the ring bearer who must carry it to Mordor and destroy the Ring by tossing it into the fires of Mount Doom from which it was made.

Frodo tells the wizard Gandalf, "It's a pity Bilbo didn't kill Gollum when he had the chance." Gandalf replies, "Pity? It is pity that stayed Bilbo's hand. Don't be too eager to deal out death and judgment. Even the very wise cannot see all ends. My heart tells me that Gollum has some part to play yet, for good or ill, before this is over. The pity of Bilbo may rule the fate of many."

Then Frodo says, "I wish the Ring had never come to me. I wish none of this had happened." Gandalf replies, "So do all who live to see such times, but that is not for them to decide. All we have to decide is what to do with the time that is given to us. There are other forces at work in this world, Frodo, besides the will of evil. Bilbo was meant to find the Ring. In which case, you also were meant to have it, and that is an encouraging thought."

From the earliest days of human existence, our curiosity about the world has driven us to understand the mysteries of the universe. Central to this pursuit has been the quest to understand matter, the fundamental substance that comprises everything we see and touch. The ancient Greeks suggested that matter is composed of indivisible particles called atoms, which laid the foundation for future scientific explorations.

Much later, the advent of quantum mechanics in the early twentieth century challenged our classical understanding of matter. Scientists developed theories that revealed the dual nature of subatomic particles, manifesting both as particles and waves. This provided a framework for understanding the behavior of subatomic particles and explained phenomena such as wave-particle duality, quantum superposition, and electron energy levels.

Numerous mysteries remain, including dark matter. The mysteries of dark matter continue to captivate scientists as they deepen our understanding of the universe. Dark matter seems to exert gravitational forces that influence the motions of physical matter, even though it doesn't interact with light. We infer the presence of dark matter through its gravitational effects on galaxies and galaxy clusters. Various hypotheses, including undiscovered particles, have been submitted in an attempt to explain dark matter.

Mass is said to be a measure of the amount of matter, but it is really a measure of stability. The larger the mass, the more momentum, or stability, it has. Science does not know what causes stability. They say everything would be rushing about at the speed of light if it were not for a mysterious cosmic molasses that slows everything down, which they call the Higgs Field.

What many people don't understand is that light does not exist in the physical universe because light does not see physical dimensions. When you observe light you witness the interplay of the non-physical with the physical. The physical too would be rushing about at the speed of light, just as the non-physical, if not for the cosmic molasses. Where does this molasses come from? What causes stability? Stability occurs because two things agree.

The famous-double slit experiment leaves us with profound implications about stability and the relationship between the observer and the observed within the realm of quantum mechanics. This experiment demonstrates the wave-partial duality of quantum objects, such as electrons or photons of light. It challenges our classical understanding of reality.

In the double-slit experiment, quantum particles are fired one by one at a barrier with two narrow slits and observed on a screen behind, they create an interference pattern similar to what is expected of a wave. This pattern suggests that each particle behaves like a wave that passes through both slits at the same time, interfering with itself and creating an interference pattern of light and dark bands on the screen behind.

However, when an observer or measurement apparatus is present, something remarkable happens. The simple act of observation collapses the wave function, causing the particle to be detected as if it had only passed through one of the two slits. This wave function collapse eliminates the interference pattern, as the particles now exhibit particle-like behavior, leaving two very distinct bands on the screen behind.

This raises intriguing questions about the relationship between the observer and the observed. It suggests that the act of observation, or measurement, can influence the behavior of quantum particles, shifting them from a state of superposition to a state of definition.

Stability occurs when two things agree.

Matter with a very small mass, such as a quantum particle, can be very unstable. Quantum particles can appear out of nothing and vanish. When you look at them they are there. When you look away they disappear. In other words, when you provide coherency by agreeing that they should be there, they are there. Once you decide to no longer look at them, then the coherency is removed and they are no longer there. They again retreat into nothingness once their job is done. They are removed after they fulfill the purpose of their creation.

Our sun is a star. All stars are suns. There are different types of stars. The star's type is determined by its mass and the phase of its life cycle. Our sun is said to be an average star with an average mass that is about halfway through its life cycle. It is said to be a yellow star due to the spectrum of light that it produces. You may think that an object with a very large mass such as a star would be stable. However, stars can be very unstable. The greater the mass of a star, the more unstable the star becomes. Very massive stars can explode in supernovas.

Our sun appears to be formed by a gravitational conglomeration of matter. You can say it was born when it first produced light. Our sun is presently in a balanced duality of heat and gravity. As heat is generated, mass is released. Eventually, mass is released until heat dominates and it expands to become a red giant star. After more heat is released then gravity dominates and it becomes a white-dwarf star. Eventually, even the white dwarf star will die. It will be removed after it fulfills the purpose of its creation.

Earth was born and it changes. Earthquakes shift continents. Mountains rise only to be flattened. Shifting tectonic plates cause earthquakes and volcanos. Climate change shifts deserts and jungles. Vast ice sheets advance and retreat leaving behind fertile soil. Climate change continues. Climate change rattles people and forces them to change.

The Earth is also gradually losing its moon. The Moon affects the length of a day and the precession of the Earth. Without the stabilizing influence of the moon, life on Earth would radically change. These effects will not become significant until millions of years pass and who knows where humankind will be then. Regardless of planetary wobble, it appears that the Earth must follow the fate of the sun. It will be removed after it fulfills the purpose of its creation.

Science teaches that all matter may be unstable and that it all may eventually decay and disappear into nothingness. If this is true, then it can be said that all matter is removed after it fulfills the purpose of its creation. It is removed when two things no longer agree. It is removed when the dream of duality ends and there is only unity.

Your body was born. It is always in a process of change and it will eventually die. It will die after it fulfills the purpose of its creation. The purpose of its creation is to experience the dream of duality.

During the dream of duality, humans came into possession of matter and it drove them to murder. It consumed them and turned them into a pitiful creature with a divided mind called the ego that both loves and hates matter. The ego hates matter for what it has done to it, but it has identified with physical forms, and now it can't bear to be without them. Through identification, it has become matter. It now loves and hates itself. It holds the object of its desire, and its malice, in its hand, and obsessively repeats, "Precious, precious . . ."

The pain body is a history of ego, which is a history of control, judgment, and cruelty. It is the latent pain of the separation and has a will of its own. For the pain body to survive it must have the ego, which believes in opposition. For the ego to survive it must have the pain body, which serves as a reservoir of pain. You will continue to be self-possessed in time, matter, and the pain body, as long as you believe in opposition and fail to forgive the past.

The pain body is the Ring of Power that dwells in shadow and awaits an opportunity to possess an ego and drive it to madness. It will cause the ego to say, "I must have the Ring of Power so that I can fight my enemy." But what enemies do you have? If you fight old age, then one day you and your enemy will meet, and what will you say to him then? If you fight death, then one day you and your enemy will meet, and what will you say to him then? When you meet your enemy, will you still identify with physical forms to defeat the opposition, or will you remove pain and become perfect Love?

War is the product of identification with physical form, but war also serves as a poignant demonstration of the transitory nature of physical form. Thus, the pain of war can bring awareness to the folly of the ego. The ego identifies with matter because it believes this brings love and security. It does not know that it identifies with matter because it is empty and blames its emptiness on an endless foray of opposition. The ego does not know that it makes its own reality. It does not know that security only appears when you give it away. It does not know that Love only appears when you give it away. It cannot give security and Love because it can only project itself. Therefore, it can only project emptiness.

The deeper you dive into the physical reality of matter, the more it slips through your fingers and disappears into an expression of possibilities. What is the purpose of this expression? The answer to this question is incredibly simple, yet profound.

The purpose is for you to become Grace. Grace does not see pain. The ego will see a history of pain, including events such as World War I, haunted by the faces in the water. The ego will say, "I wish these faces had never come to me. I wish none of this had happened." But these faces will drive you to the Truth. You were meant to see the faces, and that is an encouraging thought.

CHAPTER 36

The Sword in the Stone

The Story of King Arthur gives us the sword in the stone. Strong, powerful knights ambitious to become king had attempted to pull the sword from the stone to no avail. The sword was removed by the young boy named Arthur of seemingly little strength. How could the sword be removed by a feeble boy when it could not be removed by a powerful knight? The answer is that the boy did not remove the sword alone. It was removed by the boy *and* the stone, together.

You haven't removed a sword from a stone, but you have found stones and tried to throw them. If you throw a stone, you may say that you are the subject and the stone is the object. But in Eternity there can be no boundary between subject and object, and if there is no boundary, then who is doing the throwing? Are you throwing the stone or is the stone throwing you? The truth is both you *and* the stone are doing the throwing. You and the stone are a duality. This duality creates the illusion of throwing.

The illusion of throwing makes it possible for you to throw a ball. This makes it possible for you to play baseball. This makes it possible for some to play baseball and be well compensated as a professional. This makes it possible to have a nice house, a nice car, and free time during the off-season to do as you please. This makes it possible to have an abundant Life with freedom, family, friends, and enjoyment. This makes it possible for you to make the world a better place through your generosity and graciousness.

There is no boundary between illusion and non-illusion. In the example of the baseball player, the act of throwing is an illusion and so are the baseball, the stadium, the money, the house, the car, the freedom, the family, the friends, and the enjoyment. It is all an illusion. You may say that family and friends could not possibly be an illusion, but in many ways, they are an illusion. Their bodies are an illusion. Their individuality, to a large degree, is an illusion.

The ego claims to be the subject but is afraid of being the object. Therefore, through identification, it seeks to make all things part of the ego so it will never submit to being the object. Therefore, the ego can never trust, can never Love, and can never find harmony in a shared relationship.

You are an object because you are an object of creation. However, you are created to be the subject, and to become the subject you must remove fear limitation. To remove the fear limitation, you must create. To create, you must give. To give, you must see your opposite. To see your opposite, you must be your opposite. To be your opposite, you must be both subject and object. To be both subject and object, you must be one with Life.

If you believe that Life begins at birth, then you believe that experience must begin. You place a boundary in Eternity between experience and no experience.

If you believe that Life ends at death, then you believe that experience must end. You place a boundary in Eternity between experience and no experience.

If you believe that Eternal Life begins after death, then you believe that Eternal Life must begin. You place a boundary in Eternity between Eternal Life and no Eternal Life.

All is Eternity. All is Life. All is Eternal Life

Eternity is either all dead or all life. If you say that Eternity is all dead, then you must say that all transcendent experiences are an illusion, an illusion experienced by dead matter. This may seem reasonable, but this position places a boundary in Eternity. There can be no boundary between illusion and dead matter. Dead matter must therefore be an illusion, and therefore you should agree with me after all.

If you try to be clever and say that God must be an illusion, then you must say that you too are an illusion, but this is actually getting close to the truth. You are an illusion because you don't realize who you really are. You are the expression and experience of an illusion that is perfect for you. It is perfect for you because otherwise there would be a boundary between perfect and imperfect.

If you wish to place boundaries in Eternity, then that is your choice. As you place your boundaries, I invite you to look within and ask who is doing the placement. I trust as you do so, you will find something more than dead. I trust that you will find Life. If you find only dead matter, then I still claim that there is no boundary between you and me. Together, we have Eternal Life.

CHAPTER 37

Space

S pace is not empty. Space is Life.

The Western mind is deductive. It sees details. It sees edges. It sees the trees. It sees matter. It seeks to control. It seeks to control matter.

The Eastern mind is intuitive. It sees vastness. It sees smoothness. It sees forests. It sees Space. It seeks harmony. It seeks to harmonize with Space.

The Eastern mind also recognizes matter. The Western mind must learn to recognize Space.

Space is not just something that surrounds the planet; it also surrounds you now. We do not exist within the context of objects. We exist within the context of Space.

The Western mind should sit in a room and contemplate what exists between the walls of the room. What exists between the walls is Space. But what is this Space? In what context does this Space exist? In what context do the walls exist? In what context does the room exist? In what context do you exist?

I encourage you to stand outside on a dark night, look up at the sky, and contemplate the vastness of Space. As you contemplate, push deeper into the vastness. Consider the context in which the vastness exists. "The vastness beyond the vastness," as I like to call it.

Soon you may be unable to stand. Your knees are weakened. Your mind cannot comprehend a context that would allow you to exist. You cower because your very existence feels threatened. It is not the higher self that feels threatened. It is the lower self that feels threatened. What you feel is the conflict that exists between ego and Eternity. The ego fears Eternity.

When you are not able to stand and fall to the Earth in fear, you feel humbled by Eternity. You may ask some big questions: What is the context of my existence? Why does Eternity overwhelm me? How can I move beyond a state of humility so I can stand in Eternity's presence?

You might later find yourself saying that it is a miracle that anything can exist, that the context seems to be a miracle. You might begin to believe in miracles. You might believe in the miracle that one day your questions will be answered.

As you receive your answers, you may start appreciating Life so much that your Life seems to be perfect. You may find yourself saying that even if you die now, and it all faded to black, it would all remain perfection. You may find yourself saying that you Love God more than yourself and that you Love all people as yourself. You may then be prepared to die before you die.

One day you will be introduced to the One Life. The veil that is boundary between you and Eternity will be ripped out from underneath you as it falls crumpled on the floor. Then you will see that Truth is the context in which Eternity exists. Then you will say that Truth is everywhere and everywhere is Truth. Then you will look up at the sky and hear laughter, for all is the smiling face of Truth.

CHAPTER 38

A Den of Thieves

There are those who say the success of planet Earth is in doubt, but I say that the corner has been turned, thanks in large part to Mikhail Gorbachev and the Soviet Union choosing the path of peace, and we are now moving into solution. The divisions within the children of Abraham will dissolve because they come to recognize the One Life, and they will become aware of both the demon within and the nature of the anti-Christ.

Western religions will come to realize that the remaining opposition is not other religions, rather it is an expression of control that has no intention whatsoever of serving the Light. The children of Abraham will help the anti-Christ realize that it, too, has been deceived, and as it has lurked in shadow and cunningly conspired to build the final great empire, it has succeeded in doing nothing more than plant the seeds of its own destruction.

The Earth is a temple being sold at a price. Who is doing the selling? That would be someone who believes they own it. That would be someone who believes they own all the kingdoms of the Earth. That would be someone who sees you as nothing more than an animal to be controlled. That would be someone who only sees themselves as an animal because that is the only reality they know. That would be someone called the demon.

At the global level, the demon is called the ant-Christ. The first anti-Christ is ignorance, obedience, and slavery. Obedience and slavery create the second anti-Christ, which is anger, empire, and colonialism. However, empire and colonialism encircled the globe and eventually removed ignorance, obedience, and slavery. Thus, the first anti-Christ created the second anti-Christ, but the second anti-Christ served to destroy the first anti-Christ.

The second anti-Christ created the third anti-Christ, but the third anti-Christ served to defeat the second anti-Christ. The third anti-Christ now stands as a demon more powerful and cunning than the previous two. It is a monster from the darkest depths of hell and still largely remains in shadow. The third anti-Christ is the bottomless pit of greed and debt.

The third anti-Christ spins wealth out of smoke and vapor and then offers you the gift of wealth in the form of debt. To release yourself from toil, you accept the gift of wealth. As you do so, you surrender some of your freedom. One day, the gift of wealth is revealed to be nothing but smoke and vapor, and you have fear. In your fear, you surrender even more freedom to obtain the safety of the herd.

The third anti-Christ seeks to control the whole Earth and create a new world order. It seeks to build empire and enslave beyond any previous anti-Christ. It perceives itself to be all-powerful and in its mind it owns all the kingdoms of the Earth. It perceives itself to be the puppeteer and you are the puppet. But the third anti-Christ is deceived. It looks in the ego mirror when it sees a puppet.

The first two anti-Christs sought control, but they brought Oneness to the Earth. The third anti-Christ serves the same purpose. It will bring Oneness to the Earth as it creates the final great empire.

The third anti-Christ exists in the kingdom of possibilities. One possibility is that it moves out of shadow and openly and honestly promotes Oneness. Another possibility is that it remains in shadow and continues the path of deception and control. If it remains in shadow, then Life will remain its enemy. It will find itself on a Masada as it attempts to protect itself from the unexpected approaching at lightning speed.

The anti-Christ is a karmic cycle of obedience, empire, and debt. If you are obedient, then you build an empire, and the empire becomes debt. If you become debt, then you must again have obedience to pay the debt, and the cycle repeats.

In recent Earth, history empires created debt and debt attempted to return the cycle to obedience, but failed to do so. Debt failed because there is freedom. There is freedom, because those with Truth stand before the smoke and flame, and say to the demon, "You shall not pass!"

CHAPTER 39

The Fall

M odern reality gives us the world of Newton, Darwin, Freud, Einstein, the Hubble Telescope, nuclear reactors, and cell phones. Stories of the Cosmos are based on a modern scientific perspective that gives you a reasonable understanding of rainbows, moons, and stars. Even in these modern times, religion may present you with highly improbable stories that are said to be relevant to your salvation.

People who lived thousands of years ago had their own reality. They produced stories based on their own perspectives. Is there any truth in their stories? Does anyone with a modern reality really need to accept that creation was a singular event? Must you accept that Adam was created out of dust, basically in a one-step process? Must you accept that all the animals of the Earth congregated in pairs to board an ark? Must you accept that God placed his bow, his weapon, in the sky as a sign that the waters shall not rise again? Is it God's bow, or is it light's manifestation, a rainbow?

Those who bestow absolute authority on the Bible may say that you must believe such stories to have faith. As I have said, faith may be an obstacle on the spiritual path, which is about honesty, not faith. It doesn't matter if all the tales of the Old Testament are factual. What matters is that they are shadows of your existence—they're your possibilities.

For many centuries, theologians and historians have struggled since the early days of Christianity to explain why the God of the Old Testament is a capricious brute, while the God of the New Testament is a God of Love. The simple answer is that the Bible, like the Earth, is an expression of a human being. A human being experiences a fallen state embodying an ego that projects its capricious, brutish nature onto an idol. When the human being awakens, the ego is removed and they see the God of Love.

Christianity contends that the fall spoiled God's plan and is therefore unintentional. Mormons say that the fall is part of God's perfect plan and is therefore intentional. Like most opposing points of view, the higher perspective lies somewhere in the middle.

The fall cannot be unintentional, because that implies a break in reality. The fall must therefore be intentional as Mormons claim. It is also the product of deception, which means that someone was being intentionally deceived.

Mormons will give you intentional deception. It places you in a testing ground of free agency where you choose the right over the wrong so that you might be promoted to the next step of eternal progression. If this is true, then why did Jesus say as he hung on the cross, "Forgive them for they know not what they do?"

Mormons answer by saying that the people who crucified Jesus were simply ignorant, as Jesus says. If ignorance brings forgiveness, must we conclude that under free agency, ignorance relieves responsibility and is therefore a good strategy for receiving forgiveness?

Mormons may answer by saying that we cannot intentionally remain ignorant and that we are responsible for making the best choices we can, given our best understanding of what is right and wrong. If this is true, then how does anyone determine accountability because everyone's Life experience and level of ignorance is unique? Mormons may explain that we are all born with the light of Christ to guide us and that Christ has the power to judge each person individually.

This has brought us back to an important concept: the individual. Those who believe in free agency have a belief system that reinforces the concept of individuality, not Oneness. They believe that you are achieving salvation through a process of best effort choices on a playing field of opposition and the exact nature of this playing field is unique to you, but you must choose well or perish.

Mormons interject and say that they don't reinforce individuality because they strongly believe that salvation is only found as a family unit. Therefore, we should restate this belief in terms of a group. They believe that a group is achieving salvation through a process of best effort choices on a playing field of opposition and the exact nature of this playing field is unique to the group, but the group must choose well or perish.

Does this sound like a plan of salvation or does it sound like a tribe of prehistoric people trying to survive in a lone and dreary world? They don't recognize it, but those who believe in free agency are egos that have succeeded in equating salvation with survival.

Mormons may counter by saying that salvation for you or the group is not a matter of best-effort choices because salvation is only achieved by obedience to the laws and ordinances of the gospel. However, all this does is bring us back to what Jesus said as he hung on the cross.

The ego is a false expression of individuality that manifests at both the personal level and the group level as a collective ego. The collective ego will pretend that it promotes oneness and brotherly love, but will always say that you, or your group, are on a testing ground of opposition because the ego *is* belief in opposition, which is what caused the fall in the first place.

Religious groups believe that free agency is a prep school offering their graduates high placement in a future reality, only the future reality never arrives. It never arrives because the ego does not accept that happiness is only found in the present moment and that salvation is Now. There is free agency, but only in the dimension of Truth. In the dimension of Truth, you choose to imprison yourself in the human dimension, and this encasement becomes the ego, the very caretaker of deception. The Truth now seeks to unveil the deception and awaken you to your true nature, which is one of freedom and perfection, by placing you in a classroom where your fear limitation becomes Grace.

You may believe that you would never imprison yourself, but that is part of the deception. The deception tells you that you are an individual with free agency who is capable of independent action, but what action do you take? You call yourself a victim of God's spoiled plan, chance, or person in disagreement with you.

You can't be both independent and a victim because victims are tossed by external forces, but the independent are not. Therefore, you have deceived yourself by placing yourself in a contradiction. To remove the contradiction, you must recognize that you are not a victim, you are not an individual. You are the One Life.

Where does this leave your free agency? It leaves it right in your lap waiting to be claimed, but perhaps it remains hidden from you. When you are no longer deceived, you will have your Truth, you will see perfection, and you will be free.

If you are a Christian, then you may not be free. You are deceived by the boundary you place between God's plan and God's spoiled plan. There never was God's spoiled plan because all is perfection.

If you are a Christian, a Mormon, or a Muslim, then you may not be free. You deceive yourself by placing a boundary between who will be redeemed and who will not. God's perfect plan does not allow you to fail because God's perfect plan is backed by the power of the Cosmos, which you don't see.

When you transcend the ego, it becomes clear that the ego is not evil; it simply doesn't know what it is doing. Therefore, it is not under free agency and it cannot sin, but it can remain in fear limitation. For the ego to have free agency, it must also have control, but control is also an illusion of the ego. If you pursue the illusion of control, then you self-destruct.

At worst, you will die a tyrant who succeeded in killing his enemies. At best, you will die a righteous man who never succeeded in redeeming himself. You may ask: How can a man redeem himself if we are redeemed by Grace? The answer is that at some point, you must recognize Grace as yourself. You are Grace.

The fallen state is the ego and the ego is deception. The word deception sounds sinister to the ego, because the ego does not wish to be bamboozled by the opposition. To the keepers of Truth, deception does not sound sinister because there is no opposition. To the keepers of Truth the fallen state is a game of make-believe.

God's power and innocence are here and Now and you will see them when you can see God's perfection here and Now. Perfection always works in your best interest. This is true at the personal level and true at the larger level.

You have tried your best to foul things up, but your ability to foul things up is not greater than the Cosmos's power to redeem. You and the Earth were redeemed before the fall because time is an illusion and all is perfection.

I recognize the Old Testament as an allegory of my fallen state, and you may recognize it as an allegory of your fallen state as well. This allegory is as follows:

You are a child in the garden of innocence amongst the wonders and timeless giants. Evil characters intrude upon the garden and you are banished to a lone and dreary world of judgment and fear.

Now you must survive in a fallen world of ignorance, anger, and greed. All may seem lost with no hope of redemption, but if you remain alive and take one more step toward a brighter day, then the greater evils that are the source of your despair are washed away as if by a cleansing flood, and a rainbow of hope is placed in your sky.

You are no longer in the garden, but you are grateful to find yourself again on solid ground as you follow your path. Life again appears to hold promise, but the promise is not easily obtained and you are sold into Egypt. In Egypt, you serve in the mud pits. You churn straw and mud to make brick for the glory of princes and kings. You place faith in the redeeming blood and are spared when death and destruction come knocking at the door. Often, those who esteemed themselves as princes and kings are not so fortunate.

You leave the mud pits and are free. The land of promise seems close at hand, but you must first sojourn under the Law and choose well in the game of cause and effect. After you have chosen well you are led to the land of promise where you carry the Ark of the Covenant, conquer all before you, and drive your enemies to near extinction. You have established yourself in the promised land with strength. How much of this strength is your own and how much is of a higher power is according to your own perceptions and choice.

You seek to protect your kingdom, but you are at the crossroads of other powerful kingdoms as well, and they too would like to expand their empire. You and others are heroes. You may be a powerful king, you may be a prophet, or you may be armed with a sling and face a giant only to drop him with a well-placed stone. Sometimes, there is one hero too many because heroes are like powerful kingdoms. They both like to have a larger piece of the empire.

You are severely tested. Sometimes you pass the test, and sometimes you don't, but in either case, you remain faithful. You may be very wise, but still, it becomes almost impossible to tell whether you are faithful to your God or faithful to an idol. There are many voices. Some voices speak of glory and others speak of harmony. Sometimes the voice of harmony asks for great sacrifice. Sometimes you are deceived. The voice of harmony is really the voice of the demon that asks for sacrifice to bring glory unto him. The voice you follow is your choice. Your choice, in turn, becomes your voice.

If you do not choose well in the game of cause and effect, then you may find your kingdom divided with half or more being removed to distant lands. You may even find yourself again in captivity trying to win your escape-to return to a life of harmony, in the land of milk and honey.

In this captivity, you begin to see that a Life of harmony is beyond your reach because you see evil and fear conquer even champions of the game of cause and effect. You are told that success in this game comes by obedience, yet obedience presents you with a paradox: If I am good because I obey, then what makes me obey? If I obey because I am good, then what makes me good?

You remain divided. On one hand, you recognize the paradox, but on the other hand, your desire to return to a life of harmony reaffirms the importance of the Law, which you were commanded to obey. You attempt to resolve the division with a renewed and fervent dedication to the task of obedience, but you find that this does not bring you to harmony because no matter how much you obey, you are still not free. You realize you have been deceived because the game of cause and effect cannot be won beyond stalemate. You search for a higher game where the paradox of obedience does not exist. You search for a game that can be won.

You search for a Messiah.

CHAPTER 40

The Messiah

The Hebrew Messiah concept developed out of the realization that Israel is failing. Israel believed in their God, Yahweh, but they remained polytheistic in both practice and belief. They believed in other Gods including the Gods of other nations. They perceived a contest of nations as a contest of the Gods. Their God had failed them when they found themselves divided, scattered, conquered, and exiled, with their Temple destroyed.

To explain this painful experience and not diminish their God, their failings became part of their God's big plan. For God's big plan to exist, the God of Israel had to be big enough to encompass a plan that included all nations. Therefore, the God of Israel became the God of all. The God of all had allowed them to fail because they had worshipped idols, so the story goes.

This is how monotheism was born into Western civilization and what inspired stories of the coming Messiah. Contrary to popular belief and teaching, It is not a story based on explicit revelation from the beginning. It is a story that developed over time as the Israelites suffered and came to recognize that the painful mess in which they found themselves was due to idol worship, but there is also a higher solution, other than the Law.

So it was with Israel. So it is with you. You suffer, and you have your burdens, and when it is time to face the unexpected, you will find that you have identified with an idol of physical forms and will be looking for a higher solution, other than the hounds of judgment.

People worship the idol that they judge to be most suitable for defeating the opposition and the opposition is always another idol. You too are an idol that must be preserved, and this idol is called the ego. The ego is an idol that worships idols. The ego is not you, because it is the false self that identifies with physical forms.

As people continue to identify with physical forms, they believe it is the role of the Messiah to preserve physical forms such as their nation, their religion, their body, or the bodies of their family and friends. But as the southern kingdom of Israel found out in Babylon, and as Nando found out on the mountain, it is not the role of a Messiah to save you in your state of idol worship; it is the role of a Messiah to reinvent your sense of self.

Moral codes don't save because they don't reinvent your sense of self. Your sense of self, often an ego, and any moral code, is a projection of the ego. The ego is in constant denial of your innate goodness, and the moral code will also deny your innate goodness. Israel received a moral code called the Law, which provides guidance, but it also contained institutionalized judgment and guilt. That is why the apostle Paul said, "I was alive without the Law once, but when the commandment came, sin revived, and I died."

Even though sin was revived he still called the Law good and said, "Was then that which is good made death unto me? God forbid. But sin, that it might appear sin, working death in me by that which is good; that sin by the commandment might become exceedingly sinful."

In other words, judgment not only brought the sin but also amplified the sin by creating a contrast between what is good and what is bad. He also said, "To will is present with me, but how to perform that which is good I find not. For the good that I would, I do not; but the evil which I would not, that I do." In other words, he had a desire for good, but he had a divided mind that was at war with itself.

The divided mind in a state of ego that seeks to control all things, including lusts and passions, is unstable as concurred by James, who teaches, "A double-minded man is unstable in all his ways."

Paul goes on to say, "The carnal mind is enmity against God, for it is not subject to the Law of God, neither indeed can be." In other words, the ego, or separated man, has a false reality and does not know what it is doing.

Paul then says, "But ye are not in the flesh, but in the Spirit, if so be that the Spirit of God dwells in you . . . For as many as are led by the Spirit of God, they are the sons of God . . . and joint-heirs with Christ; if so be that we suffer with him, that we may also be glorified together."

In other words, when you transcend the ego, Consciousness dwells in you and you become a Son of God who serves with Jesus to redeem the world.

Many Christians believe that it is not by works, but by Grace that we are saved. They say that Grace provides a vicarious sacrifice, which overcomes the demands of justice, and thereby cleanses them of sin. However, there is no sin, there is no vicarious sacrifice, and there are no demands of justice. There is only fear limitation, service, and perfect Love.

It is perfect Love that redeems the world.

If you are a person of the book then you likely have a Westernized mind that sees boundary and separation. You often fail to recognize unity and therefore experience this Life as an individual claiming choice and control. Your concepts of Grace, sacrifice, justice, and sin are based on absolutes—absolute guilt, absolute loss, and an absolute boundary between you and the divine.

The Westernized mind says that man is made in the image of God and that man has an indwelling spirit, but man is not divine. Man is effectively presented as an external object that becomes a play toy of God, and, in turn, God, and even Jesus, ever remain external objects to man. In this belief system, man may be saved, or just as readily tossed on the trash heap.

Who is it that values people conditionally and tosses them on the trash heap? That would be the ego that makes alliances to defeat the opposition. The Western God is a projection of the ego that has made an enemy of God and therefore denies the divine within. There can be no boundary between you and the divine; therefore, you are the divine.

Idol worship has its place, but at some point, you must remove the contradiction. Christianity says that Jesus must be worshipped. They then say that Jesus wishes to have a relationship with you. If these are both true, then we must conclude that Jesus desires a relationship of worship, which is counter to everything he ever taught or exemplified.

In Zen Buddhism, they say that if you see Buddha walking down the road, kill him. What does this mean? It means that you must remove all identification, even identification with a person you consider a guide or guru. It even means you must remove identification with your Savior.

This will come as a shock to Christians who wash the feet of Jesus with their tears, and I know many of them do. If it is a shock, then you are not at peace. If you are not at peace, then you have not seen the dimension of Truth. If you saw Truth, then you would see Jesus as perfect Love, you would not judge, you would forgive, you would recognize all as perfection, and say there is no sin. You would say the demands for sacrifice are an expression of the ego's pain.

The ego does not know what Love is. It does not know who Jesus is. If you have not transcended the human dimension, then it is impossible for you to believe in Jesus because you don't know who Jesus is. All you have is a mental construct that you are identified with.

This identification is not honest in that it contains both truth and falseness. It is part of a false reality. The best you can do is worship an idle. This is why some who claim to be a believer in the transcendent Jesus may also be the first ones to reject a message of transcendence brought back to them from the dimension of Truth.

In the New Testament, Jesus tells you to be perfect. However, this is a mistranslation because, in the original Greek, Jesus tells you to be whole. Either way, if you are to be perfect or whole, then why do you require a Savior? Why do you require someone to sacrifice for you? Why do you require someone to defeat the opposition?

I have been taught that you should release control and give to Life. To truly give something you can't expect anything in return. When was the last time you truly gave something? Did you give your life to Jesus? What did you expect in return when you gave your Life to Jesus? If you expect to be saved, then you did not give your Life, but instead, you have made a deal. Therefore, you are an ego who has formed an alliance to defeat the opposition of death and damnation.

Jesus tells you, "No one comes to me unless the Father who sent me draws him." How does the Father draw the ego, which the Calvinists call total depravity, unto him? You are drawn to Grace by Grace, but subtle is the Lord, because if you are professing a belief in Jesus, then you might be making a deal, and if you are making a deal, then that is reinforcing the depravity.

If you say, "I am broken, save me, Jesus!" Then the question is: Save you for what purpose? So you can escape your fears rather than face them? So you can satisfy your bottomless wanting? If you say, "I am broken, fix me, Jesus!" Then this is much the same as asking for Truth because it places very little preconditions on Truth. You are recognizing that your reality is presently painful and you would like the pain to be removed, which Grace is more than happy to do.

The only precondition this request places on Truth is your preconceived story of who Jesus is. Jesus is Grace, but Grace does not save you. Grace shows that you are not an ego who is separate from other people. Therefore, to weaken the ego, Grace may wish for you to spend time with people of other belief systems as part of your rehabilitation program. Rehabilitation requires that you leave the bunker. Leaving the bunker is how the Father draws you unto Jesus. You might say the Father draws you unto a new reality that is free of pain.

In the Greek, Jesus identified himself as "a Son of God," not "*the* Son of God." He told the Jews, "Why do you say that I cannot be a Son of God when your scriptures say that we are Gods." There is no boundary between you and God. The divinity in Jesus is the divinity in you. He is Truth and you are Truth. He is a Son of God and you are a Son or Daughter of God. He is the power and the innocence of the Cosmos and you are the power and innocence of the Cosmos. There is nothing about him that you cannot attain.

Christianity calls Jesus the mediator but tells us that a boundary exists between the mediator and everyone else. They say that he is divine, but no one else. They say that he was born of a virgin, but no one else. They say that his mother is a blessed mother, but no one else. They say that he was begotten of the Father, but no one else. They say that he suffers for the sins of humanity, but no one else. They say that he and the Father are One, but no one else. They say that he comprehends your pain, but no one else.

They present a mediator that is alien to us, yet we are to take up the cross and follow him so that we can go to Heaven and sing eternal praise to another alien being, who they call God, the Father.

This teaching is contrary to the Law of Attraction, which says to follow Jesus is to be Jesus. You cannot be Jesus if you say he is divine and you are not because when you say this, you place a boundary in Eternity. This is the boundary that Christianity believes in and perpetuates. You may claim to be born again, but when you believe in boundary you will likely not see the Father, for the Father contains no boundary. The Father is the One Life that shares Eternity with Jesus. The Father is the One Life that shares Eternity with you.

I was always taught by my religion that I am a child of God and that I share a common spiritual pedigree with Jesus. However, this remained a foreign concept to me, because Consciousness is foreign to the ego. As I have said, the ego cannot explain itself in terms of Consciousness because it represents separation from Consciousness. It is a veil between you and Truth.

When you receive Truth, the veil is removed and you see clearly that you are an expression and experience of Consciousness, and *boom!* You become a Son of God! When this happens you say to the people around you, "I've discovered a whole new reality that I didn't know existed. So this is what Jesus meant by born again. I previously had no idea!"

In the East, they say there is only the present moment. They say you do not have control, so let go of it. They say that the universe guides you and that there are no accidents. It is no accident that a Messiah is placed in your universe to guide you in moving beyond pain to transcendence. A Messiah must be a transcendent Being because they must be Consciousness that reveals itself to you and thereby removes the pain of separation that fails to recognize Consciousness.

Jesus is a Messiah and he wishes to share his Life with you. For you to share the Life of Jesus, be of one mind with him. To be of one mind, have the experience and expression of a Messiah. Therefore, there are many Messiahs.

You may feel that I diminish Jesus when I say that there are many Messiahs. Jesus is perfect Love. Jesus is Truth. I cannot diminish perfect Love or Truth because neither can be diminished. If you feel that I am diminishing Jesus, then you are an ego that is identified with an idol that *can* be diminished.

You diminish yourself when you say that there can never be many Messiahs, you preclude yourself from ever becoming one. You therefore remain in your fear limitation. You become the owner of limitations that you claim for yourself. Truth does not place limitations on you but gives you problems to solve. When you solve a problem, you remove a limitation. If you claim limitations, then that is a problem, and Truth will continue to give you problems to solve until you let go of your limitations.

Your Messiah will tell you to surrender by giving up identification with all that is false. Your Messiah will tell you to forgive because guilt is false. Your Messiah will tell you to give because loss is false. Your Messiah will tell you that to conquer death you only have to die because death is false. Your Messiah will tell you to serve because you are Truth.

The vicarious sacrifice for sin does not exist, so why was Jesus crucified? This question has more than one answer, but you must begin by understanding Jesus, and to understand Jesus, you must begin by understanding yourself. To understand anything, you must perceive it and then place it in context, and to do so, you must find common ground. Therefore, let's begin by seeking common ground.

If you say Jesus is a Messiah, then you and I agree.

If you say that Jesus is the Son that has always been with the Father and he has come to suffer pain to move this existence beyond pain, then you and I agree.

If you say that the body of Jesus is the physical form of the Father that experiences mortality, then you and I agree.

If you say that the Father Loves the Son and the Son Loves the Father, then you and I agree.

If you say that when you have seen the Son, you have also seen the Father, then you and I agree.

If you say that the Father and the Son are One, then you and I agree.

If you say that the One Life and you are One, then you and I agree.

If you say that when we see you, we are seeing the One Life, then you and I agree.

If you say that the One Life Loves you, and you Love the One Life, then you and I agree.

If you say that your body is a physical form of the One Life that experiences mortality, then you and I agree.

If you say that you have always accompanied the One Life, and you have come to suffer pain to transmute pain into Consciousness, then you and I agree.

If you say that you are a Messiah, then you and I agree.

Jesus is vast just as the One Life is vast. There are more than enough possibilities for you to be a Messiah. That is why Jesus said, "You will be given a voice." If your voice condemns people to hell, then you have made Jesus into an idol that mirrors your ego's desire for opposition. If your voice debates over biblical interpretation and believes that you have Jesus figured out, then you have likely made him into something he is not. You have made him into an idol that mirrors your ego's desire for certainty.

We are Life and Life is the ultimate Messiah.

You should not await the Messiah. You should find the Messiah within.

PART 3

Millennium

Time goes by winters pass
and prosperity still lasts
Everywhere, there is peace and Love
Underneath endless skies
harmony in nature's eyes
Eagles soar
on our faith and hope knowing . . .

While we watch rivers flow
and our gifts all seem to grow
There's enough for each of us to share
Everyday music plays
singing songs of ancient ways
This we trust
The Tree of Life bears fruit for us

We stand strong together
We are all one tribe
Let this last forever
The enlightened time
Living in
The enlightened time...

(Jana Mashonee, "The Enlightened Time")

CHAPTER 41

Witness

If I tell you that Life is a game, what thought or emotion does that invoke? Does it cause you to feel liberated or does it cause you to feel threatened? Who is it that feels threatened? It is the ego.

The ego takes Life seriously. It sees Life as a vehicle to get somewhere else, and if you stir the pot by saying Life is just a game, then the ego feels doomed. You are not doomed, rather you are here, and Life is a place where you chose your reality now. How this plays out is between you and the Genie, but however it plays out, it will be perfect for you.

To emulate Jesus is not to try to comply with his teachings within the false reality of the ego; it is to awaken to Consciousness, which is something that is being universally experienced by ever-greater numbers of people as the planet continues to awaken. In my awakened state, I see no difference between the teachings of Jesus and the teachings of Buddhism. Truth is everywhere and it waits to be claimed. It is claimed when you ask for the Truth.

When Siddhartha Gautama, the Buddha, sat under the Bodhi tree he asked for Truth, he said, "Let all my skin and sinews and bones dry up with all the flesh and blood of my body, I welcome it. But I will not move from my spot until I obtain the supreme and final wisdom."

Soon he had a vision of Mara, Lord of Desire, who tempted him and also demanded to know who would testify of his worthiness to obtain ultimate wisdom. Buddha reached down and touched the Earth; the Earth shuddered. He said, "The Earth is my witness." The Earth is his witness because he and the Earth are One.

I claim to be awakened. I claim to be born again. I claim to be one with Krishna, Buddha, and Jesus. I claim to be one with you. People may ask, "Who is your witness?" I touch the Earth, but not as a witness. I touch the Earth to free her from the demon.

For my witness, I touch the stars because the stars and I are one, and I bring them within your reach.

And when stars touch the Earth, the Earth again will shudder.

Science has sought a Unified Theory of everything that can express all reality in a simple equation, but some scientists say it may be beyond an equation, that it might just be a concept or an idea.

John Archibald Wheeler, the discoverer of the quantum foam, said, "To my mind, there must be at the bottom of it all, an utterly, not equation, not an utterly simple equation, but an utterly simple idea. And to me, that idea, when we finally discover it, will be so compelling, so inevitable, so beautiful, that we will all say to each other, Oh how could it ever been otherwise?"

There is an equation to explain all reality, but as Wheeler expected, it is also an idea. The symbol I use for you is U. The symbol for Eternity is ∞. You are your own Eternity. The equation of this idea, that explains all reality is simply $U=\infty$.

CHAPTER 42

Peaceful Warrior

The Millennium is a personal experience that begins when you awaken to Truth. It is the enlightened time where there is peace and Love. You experience the mind's true liberation and become a portal for Consciousness to flow into this world so that suffering may end as Earth enters a state of Being and communes with Truth.

The Millennium is also known as the Great Awakening, the Messianic Age, and the Age of Aquarius. It has been described in modern times as an age of harmony and understanding, with sympathy and trust abounding, and no more falsehoods or derisions.

It has been described by the Jews of medieval times as an age when there will be no hunger or war, no jealousy or rivalry, and the entire occupation of the Earth will be only to perceive the esoteric wisdom of God. As written in Isaiah during the time of Assyrian conquest, it is an age when the Earth shall be filled with the knowledge of God as the waters cover the sea.

As a Zen warrior, I have contended with a gray-bearded wizard. The mere mention of this wizard's name would erase the courage and befuddle the minds of my more fearful predecessors. The wizard possesses great vision and can anticipate your every move. I led my army into the wilderness to battle the wizard but was defeated.

However, I did not retreat north as others have done in the past. Instead, I turned my army south to the heart of the wizard's kingdom.

I continued to fight the wizard, day in and day out, and would not relent as I continued to try and outflank him. I made mistakes and found myself in a stalemate. I continued to gather my forces and coordinated my efforts with other warriors.

The stalemate was broken and I pursued my enemy until I had the wizard in a death grip. We then sat down to negotiate peace. We discussed how we shared a common history in a previous conflict. I remembered him better than he remembered me. This is to be expected because I am just a common man, and he, after all, is a wizard. The spoils of war were shared with a younger man, a man of much dash and daring.

I have fought for the cause of freedom, been victorious, and there has been much cause for celebration. However, I had not yet learned the lessons of freedom. I later sent the young man off dashing and daring on a war to enslave the people of the vision quest.

These people were like the wizard; they possessed great vision and anticipated his arrival. The young man of dash and daring was not a Zen warrior; he fought for his own glory and paid a heavy price. The war of enslavement continued as a means to increase the boundary of the empire.

My experience as a slave and the unfortunate war of enslavement for which I was partially to blame made me a great champion of freedom. I now see the cause of freedom as the great crusade, and I now have little regard for the concerns of empire builders.

As a Zen warrior, I was again called to fight for freedom in distant lands where I became a master of the lightning war. My opponents were held immobilized as I struck, when and where I pleased. I no longer contended against a wizard; I contended against a madman. My victories came readily until I was confronted by my own limitations. I repented of these mistakes and again found myself in rapid pursuit of a fleeing enemy.

I have learned to love the fight for freedom and I have learned to fight for no other cause. I look across a landscape covered in glistening snow and hear the sounds of battle. It is no longer the sounds of cannon fire I hear, but the echo of trumpets, played by an angelic army, heralding a New Age of freedom.

I have been told that my warrior training is complete and that I must become a master of the darkness and a master of the light.

I have experienced the Ring of Power and the demon. I have been a ring bearer and have destroyed a Ring of Power in the fires of Mount Doom.

I now understand the workings of the Ring and the workings of the demon. I am now an active participant in the game of make-believe, and in this sense, I operate beyond the realm of human consciousness and beyond the limits of natural physical law as I dance with the rhythms of creation and play with Jesus amongst the waves.

I now have great vision. I see that the grey-bearded wizard was never my enemy. Rather, he was part of the landscape that allowed my victories to exist. I see that he is the one who reduced my second enemy into a madman. Now I am a peaceful warrior who serves the cause of freedom. The peaceful warrior does not fight, because his sword has been pounded into the plowshare of service, and the decorations on his chest have become the flowers of peace.

The above allegory of my present Earth Life is accurate. It resonates with historical events and literary symbols because I am Earth and all is loops within higher loops. In a present loop, the Earth speaks to me and brings me into the dark places to reveal the demon that is the cause of her pain. She no longer wishes to carry this heavy burden and asks to be cleansed and receive the gift of Truth. This is the purpose I serve. I have been given the power to unveil the demon and bring it out of darkness into the light.

All light is loops and loops have rhythm. We now send our light into Space with our syncopated rhythms of loops within higher loops. Our rhythms will soon be joined by other rhythms and we shall receive light. This light will serve to grant the Earth her wish.

I do not wish for you to die within the bunker. I come to remove you from the bunker so that you may have Life. To have Life, you must be free and become a friend of Life. As you do so, you will not become less. You will become more than you have ever imagined.

You are the history of Earth, and you are the future of Earth. The prophecies of Earth are prophecies that speak directly to you. Within you are the Old Testament, the New Testament, and the Millennium. Within you are the fall and the possibility of transcendence. Within you is the promise of Eternal Life.

Truth brings Love to a world of pain. All is Love when pain is removed. When all are Love, all are Truth. When all are Truth, all perceive the esoteric wisdom of God, the Earth will be filled with Truth as the waters cover the sea; all will Be the freedom and joy of the Cosmos, impossible to contain.

CHAPTER 43

Rainbows, Moons, and Stars

You are one with the One Life.

You are the eternal dreamer in the timeless present moment. You are given the gift of your own dream. In your dream is your mansion on high.

The One Life dreams of rainbows, moons, and stars.

The One Life places you within the dream.

The One Life places you amongst the stars. The One Life places you beneath the moon.

The One Life places you upon the rainbow.

The stars teach us of loops, they challenge us to move beyond our limits and sail into the abyss.

We close the loop of the rainbow and discover a New World. We learn that the abyss is an illusion. We bring Oneness to the rainbow.

The rainbow colors our minds and teaches us to move beyond our limits and discover the Oneness of light. We learn to close loops of light to color our minds. We bring Oneness to the rainbow.

The moon challenges us to move beyond our limits and venture toward the abyss. We apprentice by the light of the moon. We fly men in galleons to nearby seas.

We fly machines into the abyss, seeking to share loops of light with those of other worlds.

The light of other worlds teaches us to move beyond our limits and discover the Oneness of all existence. We will soon discover New Worlds. We will soon learn that the abyss is an illusion. We will soon bring the rainbow to the One Life.

CHAPTER 44

Apocalypse

The book of Revelation describes seven seals and the opening of each seal is the cause of some event, or series of events. The opening of the first four seals releases the four horsemen of the apocalypse.

They are usually said to represent conquest, war, famine, and death. Many interpretations of these horsemen have been offered in terms of Earth history as people attempt to link the four horsemen to events of the past and possible events of the future.

The opening of the fifth seal releases the vision of the saints and the martyrs. The opening of the sixth seal releases cosmic disturbances and heavenly signs. The opening of the seventh seal is the prelude to the seven trumpets of seven Angels and the final judgment. This releases teaching, destruction, deception, death, and a final realization of the one true God.

Armageddon is the final pre-Millennial battle where Jesus Christ comes to defeat the anti-Christ and throw him into the bottomless pit so that the Earth may exist in a state of peace and oneness with God for a thousand years. After the Millennial Age, the anti-Christ is again released from the pit, but the fires of Heaven come down to effect the final destruction of the anti-Christ.

There is no past and future, there is only Now. There is no history of Earth, there is only you. You are the Earth and you are the apocalypse. You are the seven seals; the four horsemen ride upon your landscape. You are the final battle of Armageddon. You are the final destruction of the demon. You are creation loops that exist in a universe of possibilities.

One possibility is that you enter this landscape of teaching, destruction, deception, and death. You conquer many of your fear limitations. On your path of conquest, you wage war on the demon. You win your war and then enter a time of famine where you surrender all on the altar of Truth. You then experience the death of the ego and are born again. You then share the visions of the saints and the martyrs. You see cosmic disturbances and heavenly signs in your daily existence.

There is final judgment because you see all as perfection; you cease to judge. Your creation process continues on a landscape of teaching, destruction, deception, and another death. There remains destruction and deception because you continue to change and progress in a world of perception. There is another death when you transcend to yet higher levels of Truth.

Another possibility is that you enter this landscape of teaching, destruction, deception, and death. You conquer some of your fear limitations. On your path of conquest, you wage war on the demon. In your war, you become the demon and an enemy of Life. Then the abundance of Life is withdrawn from you and you experience famine and death, because you have become death. You must then make another reckoning and wage war on the demon. You must again enter a landscape of teaching, destruction, deception, and death.

Armageddon is your final pre-Millennial battle where your Messiah comes to defeat the demon and throw him into the bottomless pit so that you may exist in a state of Being. But as I have said, when the demon is thrown down, it takes a piece of you with it. It therefore knows your weaknesses and you will again be tested. You pass the test by serving Truth, and then the fires of Heaven come down to affect the final destruction of the demon.

Everyone is on an apocalyptic path and everyone will fight the battle of Armageddon. People will try to escape them by remaining in the bunker of their faith that says only believers shall be saved. You won't be saved, but your present reality of idol worship will be destroyed.

Don't fear this destruction because it is simply part of your purpose. This book is here to guide you on your path and help you become a peaceful warrior who can overcome demons and win the battle of Armageddon. Don't fear the battle, because the battle is won with peace.

CHAPTER 45

Imagine

This book serves as a guide on your journey to Truth. This journey is your honest quest for Truth that will prepare you for the possibilities that are within you. You will thereby be prepared to meet the unexpected of this world and the unexpected of other worlds.

On my journey, I have encountered many dangers and have released many fears. I have been introduced to Truth. I have experienced the intelligence, Love, and laughter of Truth that is the Cosmos and beyond human experience. I became an egoless child, but this child also has a protector who is free of judgment. I have become the lion and the lamb, and my journey continues. I still must face many dangers and release fears. I remain on an honest quest for Truth.

Truth protects, nurtures, and teaches me. Some of the things that Truth teaches are surprising and unsettling. As I am taught and progress, it can seem like a lonely road because this experience is always personal. What you experience and express is yours alone. However, you are never alone because Truth always guides you.

My experience with Truth is full of adventure. The experience continues to build and change as it leads me to places that I have never imagined. In a sense, I am no longer human because I have transcended the human dimension and I am coupled to the dimension of Truth. I continue to explore this new dimension with wonder and amazement because that is our purpose.

As I have written this book, I have often felt the strong presence of Truth because inspiration often drives the writing process. Inspiration brings answers, but there is still exploration because Truth continues to play its game of hide-and-seek amongst the trees.

Sometimes you study other books that people give you. Sometimes you pause and think; the answers just come. Sometimes the answers come out of nowhere and you feel like you are just a hand holding a pen. I have often been inspired to write things that I did not fully understand when the words were placed on the page. I simply trust that Truth, as given at any moment, is for a purpose. That is, I trust that Truth is perfect in its place. It is my dream and my Truth that this book is perfect in its place for you.

I find myself to be a man of northern European descent who is a rebellious Mormon, who has learned Biblical Grace from a born-again Christian, and who learned of transcendence from the Wisemen of the East, sharing my personal experience of Truth as I sit and listen to inspirational music by a woman of southern African descent. This is both surprising and expected because all is the Truth.

I now find myself the author of a book written so that people may have a more abundant Life. In fact, it is a book that was created so that people may have Eternal Life. I am not only the author of a book, I have also met the Genie.

I wish for you to meet the Genie. The Genie removed all that is old and made all things new. I now live in Eternity. I now live in possibilities. When I experienced the intelligence, Love, and laughter of Truth that is the Cosmos, I also experienced myself. I am a peaceful warrior that serves the purpose of Truth.

Truth wishes for you to awaken to Krishna, Buddha, Jesus, Allah, the Father, the Mother, the Creator, the I Am, and the One Life.

Truth wishes for you to awaken, to see flowers for the very first time.

Truth wishes for you to awaken, to listen as the rocks and stones begin to sing.

Truth wishes for you to awaken, to be the Way, the Truth, and the Life.

It is time for all egos to die. It is time for all obedience to end. It is time for all empires to fall. It is time for all people to imagine a world of Truth.

> Imagine there's no heaven
> It's easy if you try
> No hell below us
> Above us only sky
>
> Imagine all the people
> Living for today
>
> Imagine there's no countries
> It isn't hard to do
> Nothing to kill or die for
> And no religion, too
>
> Imagine all the people
> Living life in peace . . .
>
> Imagine no possessions
> I wonder if you can
> No need for greed or hunger
> A brotherhood of man
>
> Imagine all the people
> Sharing all the world
>
> You, you may say I'm a dreamer
> But I'm not the only one
> I hope someday you'll join us
> And the world will live as one

(John Lennon, "Imagine")

The song "Imagine" was once something that I did not relate to. I would ask myself, "How could there be no heaven? How could there be no hell below us and above us only sky? How could there be no religion, too?" I now have the answer. The answer is that all is one and all is Truth.

One day I will fall beneath the peach blossoms. If you are a friend, then be at peace, because peaceful warriors arise and receive greater light. If you are a foe, then look within.

Within you will find the smiling face of Truth.

CHAPTER 46

EC Theory

The purpose of this theory is to bring science and religion into balance and guide us toward the next level of creation. It stands perfectly in its place of Creation. It is built upon the efforts of many scientists and religionists of today and throughout history.

This is the end product of what's been cooking in the soup pot of our civilization to date. In this soup, there are ingredients from science such as relativity, particle-wave duality, and M-theory. From religion, we have the ingredients of, universal consciousness, live matter, eternal Life, and God.

This theory takes the lid off, tastes the soup, and announces to the world that we have prepared something more delicious than we have ever imagined.

This theory shall be referred to as EC Theory, which stands for energy Consciousness or if you prefer Eternal Creation.

This theory as presented, includes little in terms of mathematics. This theory is based on wisdom.

Mathematics is knowledge. Wisdom is Truth.

Foundations

The following three postulates are the foundation of EC Theory:

Time is an illusion.

All forms exist in time and are therefore an illusion.

There is no boundary in Eternity.

Time is an illusion and does not exist. Eternity is Now. Eternity is a sequence of events, nothing more. Reality is a sequence of events. In EC Theory, concepts will be said to exist in reality rather than in time.

In Eternity, there is only perfection. There is no boundary between perfection and imperfection. All falseness becomes true when perceived from a higher perspective.

All things manifest as energy, E, or Consciousness (Cns) but they always exist in an E-Cns loop. E and Cns are a duality.

Science detects the E component of the duality or the center of the E-Cns loop. Science does not directly detect the Cns component.

E exists in three-dimensional Space. Cns exists in multi-dimensional Space. Cns is not directly detected by instrumentation. It is the hidden dimension with which our perceived dimensions form a resonant duality.

Relativity

Special Relativity teaches four principles:

- Absolute uniform motion cannot be detected.
- The speed of light is independent of the motion of the source.
- There is space-time duality.
- There is energy-matter duality.

Special Relativity is based on a model of space that is a vacuous nothingness, and that there is no boundary where the laws of physics change.

This model of space leads to the claim that uniform motion cannot be detected because there is no reference frame or boundary to measure motion. This leads to the idea that there is no boundary in Eternity.

This model of space leads relativity to claim that the consistency of the speed of light is due to time dilation.

Time dilation leads to the concept of "you can't get there from here" because your mass increases to infinity.

Special relativity time dilation has been validated by experiment. The half-life of subatomic particles has been measured to be a function of time dilation. The energy-mass duality of Special Relativity has been useful in the development of nuclear power. For example, it explains the energy states of the byproducts of nuclear fission.

However, Special Relativity begins on a foundation of no special place or boundary yet in Special Relativity there are many boundaries: maximum speed c, time dilation to zero, length dilation to zero, and infinite mass.

Thus, Special Relativity starts with a foundation of no boundary and ends up with a structure of limits, zeros, and infinities, which take on the appearance of boundaries.

In EC Theory, time is taken off the table and Space is not empty. The relativity principle of no boundary remains. In EC Theory the energy-Mass duality is replaced by energy-Consciousness duality. EC Theory is in harmony with the experimental validations of Special Relativity Theory. However, in EC Theory the limits, zeros, and infinities go away. You lose the infinities but what you find instead is Eternity.

General Relativity teaches four principles:

- Gravity is indiscernible from acceleration.
- Light, which has no mass, is affected by gravity.
- Gravity is warped Space-time.
- Time is the function of gravity or acceleration.

Observing how light bends around stars has validated General Relativity. It has also been validated by comparing zero gravity orbital clock speed with the clock speed on the surface of the Earth.

The clock orbiting the Earth onboard a satellite runs faster. For this reason, the clocks on the Global Positioning System, (GPS) satellites must be reset several times per day.

In EC Theory there is no time. Acceleration is a concept in time and therefore disappears. Gravity, as viewed in General Relativity to be a warping of Space, remains. The fact that clocks run slower in low gravity potential remains. However, gravity and acceleration are replaced by expansion and contraction of Space.

Quantum Mechanics

Quantum Mechanics teaches that energy and frequencies are quantized. It teaches that there is a fundamental minimum quantized value of energy and length. It teaches the uncertainty principle, which states that the location and momentum of a particle may only be known to a certain level of accuracy. It also teaches that particles have a particle-wave duality. They are particles in that they seem to have all-or-nothing effects. They are waves in that their influence seems to be distributed over a larger area of Space-time.

In EC Theory, energy and frequencies are not just quantized, they are digitized. Energy is transmitted and exists in digitized packets of equal amplitude. Frequency is rooted in time and therefore does not exist in EC Theory.

In EC Theory, everything exists as a sequence of events. Frequency therefore is a repetition of events that occur with a given density in the normal backdrop, or bias, of events.

In EC Theory, the wave properties of a particle are not the property of the particle; it is simply the impulse response of the observer. Thus, in EC Theory the universe is a giant computer sending streams of data bits. This data is in the form of pulses. Matter has a band-limited response to these pulses and therefore perceives them as waves.

Maxwell's Equations

Maxwell's equations express the presently accepted theory of electromagnetism. They express light as a duality of electric and magnetic fields. They also place a boundary between matter and Space. Maxwell's equations of light contain duality and boundary and are therefore only relative truth.

Here, a new theory of light is presented that points to illusions formed by higher Consciousness as it projects the human dimension. There is no boundary between illusion and non-illusion; therefore, the human dimension is all illusion.

All perceived forces and physical forms that the human mind considers separate from itself are an expression of Consciousness. The New Testament says that in the last days, the powers of heaven shall be shaken. They are shaken because they are illusions and are not really there. This prophecy is fulfilled.

Maxwell's Equations present an elegant relationship between an electric field E, and a magnetic flux density field B. Maxwell's Equations are often viewed as a unified theory, in that they unite electricity, magnetism, radio waves, and light waves into manifestations of the relationship of E and B.

From Maxwell's Equations, some may say that E and B form a duality. In Maxwell's Equations, E and B appear to form a duality in empty Space, but they are not a duality in the presence of matter. In empty Space, they say the divergence of E equals zero, and E forms closed loops. In the presence of matter, they say the divergence of E is equal to the enclosed charge and terminates on charge.

It is important to note that an E-field loop is never detected in Space because all detectors are made of matter. Radio waves are launched into Space with matter just as they are received in Space with matter. What actually happens to the E-field in Space is never perceived, it only exists in Maxwell's Equations as a mathematical model. Maxwell says that the laws of Physics are different in Space than in the presence of matter. Maxwell's Equations place a boundary between Space and matter.

Maxwell's Theory is a house built on a foundation of Gauss's Laws, Faraday's Law, and Ampere's Law. Within the walls of this house lies a boundary between Space and matter that separates us from Eternity. Maxwell's house is haunted by a ghost that has the power to remove this boundary. This ghost is named Jay Omega and he tells us, "It's time to wake up and smell the coffee!"

Maxwell's Equations can be expressed in different forms. The form addressed here will be the differential, complex time-harmonic form.

They are written in the language of mathematics as:

Gauss's Law of Electric Fields: $\nabla \bullet \varepsilon E = \rho_v$
Gauss's law of Magnetic Fields: $\nabla \bullet B = 0$
Faraday's Law: $\nabla \times E = -j\omega B$
Ampere's Law Modified by Maxwell: $\nabla \times B / \mu_o = J + j\omega \varepsilon_o E$

These equations can be expressed in common spoken language as:

- Gauss's Law of Electric Fields asserts that the electric field, E, terminates on electric charge. In the presence of matter, E must have terminations and cannot be a loop. In Space, E is a loop. Therefore, Gauss's Law teaches that there is an E boundary between Space and matter.
- Gauss's law of Magnetic Fields says that the magnetic flux density field, B, does not originate or terminate on any type of charge. B is always a loop.
- Faraday's Law says that an E loop is driven by the inverse rate of change of B. The term $j\omega$, pronounced Jay Omega, in front of B, means that the inverse rate of change of B is a phase-shifted and scaled version of B. Therefore, Jay Omega tells us that it is not B that is the causal driving force of E loops rather it is some other field that is a phase-shifted, scaled version of B. This is a ghost field, G_h. Moreover, we can say that B is a phase-shifted and scaled version of G_h.

- Ampere's Law Modified by Maxwell says that B loops are driven by moving charge (current) and the rate of change of E (displacement current). The displacement current term is Maxwell's contribution that ties electro-magnetism together to explain the propagation of radio waves and light.

In the presence of matter, B loops are driven by current and/or displacement current. In Space, B loops are only driven by displacement current. Therefore, there exists a B boundary between Space and matter.

Jay Omega in front of $\varepsilon_o E$ means that the rate of change of E is a phase-shifted and scaled version of E. Therefore, Jay Omega tells us that it is not E that is a causal driving force of B loops rather it is some other field that is a phase-shifted, scaled version of E. Let us assume for now that this is the same ghost field, G_h. Then we can say that E is a phase-shifted and scaled version of G_h as well.

Let us consider the relationship between G_h, E, and B in free Space where there is no charge or current. We can write the following equations for G_h in terms of E and B.

$$G_h = -j\omega B$$

$$G_h = j\omega \varepsilon_o \mu_o E$$

From these equations, we see that G_h leads E by $\pi/2$ and lags B by $\pi/2$. It now appears that E and B are separated in phase by π with G_h in the middle.

For two entities to form a duality they must be phase separated by $\pi/2$. This allows a balance point to exist, where both entities are at, or near, their nominal value.

Therefore E and B are not a duality in this case. For G_h to exist, a proper duality must be found involving G_h, E, and B. For now, let us assert that the duality exists in G_h alone. Therefore G_h is a resonant duality, or rather it is a G_h-Loop.

The free Space solutions to Maxwell's Equations give a plane wave with E and B in phase. The impedance ratio of the E and B of a plane wave is given as c, the speed of light. This real impedance has mistakenly been interpreted to be the plane wave delivering power.

However, it delivers power to a detector, a detector made of matter, it is not the impedance of an energy source. An energy source has a negative impedance when it is delivering energy. A phase difference of π is required between E and B for the free Space plane wave to have the correct impedance to function as an energy source. Thus, the previously accepted solutions to plane wave propagation in free Space do not satisfy the requirements for light to serve as a vehicle for energy transfer. The existence of G_h-loops points to the correct solution since it forces the concepts of E and B to take on the required phase in Space to deliver energy. This gives further evidence that E, B, or both, may not be an essential part of duality and therefore need not exist in Space.

The solutions to Maxwell's Equations in free Space have E looping back on itself and therefore forming its own boundary condition. It has E existing in a non-duality with itself. In Eternity all things must exist in duality. If one accepts the existence of E in free Space then one must accept that duality is not a requirement. Therefore EC Theory concludes that E does not exist in free Space. This removes the E boundary between Space and matter because now E is simply a manifestation of G_h duality in the presence of matter.

As was the case for E, the solution to Maxwell's Equations in free Space has B looping back on itself and, therefore, forming its own boundary condition. It has B existing in a non-duality with itself. In Eternity all things must exist in duality. If one accepts the existence of B in free Space then one must accept that duality is not a requirement. Therefore, EC Theory concludes that B does not exist in Space. This removes the B boundary between Space and matter because now B is simply a manifestation of G_h duality in the presence of matter.

EC Theory asserts Space and matter oneness. It asserts that they are manifestations of the same underlying duality. It asserts that there is no boundary between Space and matter.

EC Theory asserts that E and B do not exist in Space. But if there is no boundary, then how can they exist in the presence of matter? This question will now be addressed.

In the presence of matter, Maxwell's Equations say that E originates from a positively charged particle and terminates on a negatively charged particle. However, from Quantum Mechanics, we know that these charged particles, even when static, have wave properties themselves.

Therefore, Maxwell's Equations must say that E originates from a positively polarized wave and terminates on a negatively polarized wave. Therefore, E, in the presence of matter appears to be coupled to another wave duality. This wave duality must be the manifestation of a loop that is balanced at neutral polarization so that it may take on positive and negative polarity during its cycle.

The question is, what duality loop has these polarization properties? This duality loop is not defined in Maxwell's Equations and is therefore left as a mystery. Since we wish to remove the boundary between Space and matter, let us assume that this duality loop is a G_h-loop.

In the presence of matter the divergence, or spatial rate of change, of E is a function of charged particles in the volume. The spatial rate of change of E indicates another field. We assert that this other field is G_h-loops. Thus E is asserted to be G_h-loops, terminated, or coupled, to other G_h-loops.

In the presence of matter, Maxwell's Equations say that B-loops are formed by current and displacement current. Current is the rate at which charged particles cross the plane enclosed by the B-loop. Displacement current is the rate of change of E normal to the plane enclosed by the B-loop.

We have already asserted that charged particles are manifestations of a G_h-loop. Therefore current is also a manifestation of a G_h-loop.

Displacement current is the rate of change of E. In matter, this is rooted in the displacement of charged particles that are in a cyclic path or loop. But again charged particles are manifestations of a G_h-Loop. Therefore, displacement current is the rate of change of a G_h-loop. Furthermore, the rate of change of a G_h-loop must itself be another field or loop. There must exist another G_h-loop that phase leads the original G_h-loop by $\pi/2$. Therefore, G_h-loops form resonant dualities.

We therefore see current and displacement current as G_h-loops that exist in resonant duality with each other.

In the static case, B loops are said to be created by static current. However, there really is no such thing as a static case because current is not static. It is always charged particles moving in a loop or circle.

Acceleration is the change of speed and/or change of direction. Loop or circular motion is acceleration because the direction of motion is constantly changing as it moves around a circular path. Therefore static B is always created by acceleration.

The rate of change is synonymous with acceleration. Therefore, static B is created by the rate of change of charged particles. Charged particles are G_h-loops.

Therefore, B is created by the rate of change of G_h-loops. The rate of change of G_h-loops is G_h-loops. Therefore, static B is G_h-loops in resonant dualities.

In a non-static case let us assume that B-loops are created only by displacement current. However, we now call displacement current G_h-loops in resonant dualities.

Therefore, there is no boundary between static B and non-static B in that they are both projections of G_h-loops on a screen of matter.

In the presence of matter, we therefore have E as G_h-loops coupled to G_h-loops and B is the projection of G_h-loops.

In the presence of matter, we say that charge carriers are electrons or holes. Holes have an application to semiconductors. Electrons have applications to conductors and semiconductors so let's examine the electron and the associated E.

Electrons exist as waves within a matrix of energy levels.

The electron on the surface of a waveguide will terminate the E such that the tangential and normal E goes to zero. It is normally stated that only the tangential component goes to zero, but this describes the nature of the E in the waveguide mode, not the E-field when it terminates on an electron.

Thus, the surface of the waveguide acts like an inverting mirror that cancels E at the surface. Two conductors are brought together to form a pair of mirrors that produce an E-standing wave. As the frequency is decreased the standing wave spreads down the direction of propagation and becomes a wave bouncing back and forth between the Mirrors of the waveguide.

B also exists as a loop anchored to the spatial varying, rather than the time-varying, displacement current of the E standing wave. As the frequency is increased, E is no longer constrained by the waveguide and is said to propagate into Space. What really happens if the frequency is so high that matter can no longer resonate with the underlying duality and therefore the underlying duality exists only in Space?

In the presence of matter, E and B do not exist in duality because their interdependency involves a mingling of charged particles that are static, moving, or accelerated. These particles are not part of an E and B duality because they exist as part of matter and not E and B.

EC Theory asserts that E and B do not exist as a duality in the presence of matter. Furthermore, EC Theory asserts that they do not exist at all in Space or in the presence of matter. EC Theory asserts that E and B only exist as shadows cast by G_h-loops upon the surface of matter. The boundary between Space and matter that exists in Maxwell's house is thus removed in EC Theory.

The solution to Maxwell's Equations in free Space for E and B is really a composite of forward and reverse traveling waves. This solution would hold for G_h as well. Textbooks simply toss out the reverse traveling wave component as being insignificant. But it is significant because if you have a forward and reverse traveling wave that means you have a standing wave. But how can you have a standing wave in Space?

Space must always act as a mirror to light for a standing wave to exist. This is synchronistic with Relativity Theory, which says if you are a beam of light, then the universe shrinks to zero in the direction of travel. So we see that from the perspective of a beam of light, the universe does shrink to nothing more than two tightly Spaced mirrors.

When you are between two tightly Spaced mirrors you see Eternity in both directions.

Light must always look in the mirror and see a reflection of itself. In a waveguide, the metal surface provides the mirror. In Space, the fabric of Space provides the mirror. The essence of light must already exist in the fabric of Space.

This also removes the boundary between light and no light. Light traveling through Space must experience a shifting of the mirror. The mirror must not move else the mirror would form a boundary. Space must be a mirror. Space must everywhere consist of mirrored resonant cavities that are coupled to other mirrored resonant cavities.

The fabric of Space must be such that light is latched from one resonant loop to another resonant loop. The fabric of Space must be such that light is latched from one duality loop to another duality Loop. Each duality loop must appear as a mirror to another duality loop. Let us assume that the duality loop is G_h-loop. This G_h-loop does not exist in time, because in EC Theory there is no time. Rather they exist in sequence-Space.

In sequence-space, a clock governs events. In the transfer of light, an energy packet is clocked from one G_h-loop duality to another. There are three clock steps and three G_h-loop dualities in an energy packet transfer as shown in the table below.

Clock Cycle	Loop-1 State	Loop-2 State	Loop-3 State	Location of Energy Packet
1	0	$\Pi/2$	Π	1
2	$\Pi/2$	Π	$-\Pi/2$	2
3	Π	$-\Pi/2$	0	3

The speed at which this latching occurs determines the speed of light. It is a property of the fabric of Space. This is why the speed of light is observed to be constant. It is the speed at which the fabric of Space will deliver a packet of energy from point A to Point B.

In the study of sound waves, we know that the speed at which energy travels naturally through the atmosphere is ~ 750mph. We also know that bullets, missiles, and jet aircraft fly much faster than that.

From Maxwell's Equations, we toss out plane wave propagation in a vacuum in favor of a fabric of Space, which in EC Theory serves digital latching system that conveys energy packets.

Also, from Maxwell's Equations, we toss out E-Field and B-Field in favor of a true duality that exits in free Space and in the presence of matter. This true duality casts a shadow on the surface of matter and thus manifests E-field and B-field.

EC Theory asserts that G_h-loops are true dualities and exist in larger G_h-loop duality pairs or even larger duality sets. Furthermore, EC Theory asserts that the internal essences within a G_h-loop that form a resonant duality are energy and Consciousness.

PART IV

Gifts

You are life, for whom I dream my dream,
to give the gift of Space and time
I give you the gift so that you might Be,
to dream your dream and always live free,
and as you dream your dream, you give the gift to me
You are my life, my love, you are all that I am.
Together we dream and have space and time,
and in our dream, we never die

(Chris Grondahl, "The Gift")

In our lives, we receive many gifts. Here are some gifts that have special relevance to this book. I have also included explanations of the relevance. I encouraged you to explore these gifts on your honest quest for Truth.

The New Testament by many authors and stewards. As you read and study, look for the narrow gate. It exists in the transcendent Now. Forgive and do not judge. Love all Life and inherit the place that has been prepared for you from the foundation of the world. Separate yourself from all falseness and the illusions of the demon. Blessed are those who hunger and thirst for what is right, for they shall be filled. As you separate yourself from all falseness, you will be persecuted by the demon. Blessed are those who are persecuted in the cause of right, for theirs is the kingdom of Heaven. As the persecutions increase, you may find the Lord's Prayer to be of comfort, "Our Father, who art in Heaven, hallowed be thy name. Thy kingdom come, Thy will be done, on Earth as it is in Heaven . . . " As you progress you will be able to stand up and boldly proclaim, blessed is he that comes in the name of the Lord, for I and my Father are one and the same.

A New Earth –Awakening to Your Life's Purpose by Eckhart Tolle. My wife who heard about this book on Oprah gave it to me in early 2008. I read it in about two days. It resonated powerfully with me. Throughout my life, I have grown more aware that at some point we need to be completely free of destructive thought and emotion, or else Life will remain an endless chore. I also became more aware that we should not judge Life by labeling things and events as good or bad. Tolle taught me how to do these things. He also taught me how to leave all my baggage of the past at the station as I boarded the train to the present moment. My wife also gave me the audio CD set of this book as a Christmas gift in 2008. When I received this gift, I thought, "This is nice, but I have already read the book so I'm not sure what I'm supposed to do with the CDs." The audio CDs sat on my shelf for almost a year. In December 2009, I felt that I might benefit from A New Earth refresher course and I started listening to the CDs. One day, as I listened in my car on the way home from work, I heard him say, "Life is the dancer and you are the dance." This struck me with great power and it remains the most beautiful teaching in my Life. I awakened to the dancer a few days later.

The Power of Now by Eckhart Tolle. I read this book after I finished reading A New Earth. This book teaches that time is an illusion and that life only really exists in the present moment. This book also resonated powerfully with me. In my life, I have grown more aware that we do not place enough emphasis on the present. I noticed that people seemed to always think in terms of a future reward or happiness. I felt that we were somehow not applying enough weight to the present. I also believed from my study and questioning in physics that time is an illusion. This book also teaches that "the secret to Life is to die before you die." I died on the morning of December 19, 2009. I live in the transcendent Now.

Being in Balance by Wayne Dyer. I listened to the audio CD version of this book several times as I journeyed deeper into the garden. He quoted the Koran, "Everything evil comes from yourself, everything good comes from God." He taught me to disassociate myself from the turmoil and conflict of this world. This brought me out of judgment and deeper into acceptance, appreciation, and peace.

The Old Testament by many authors and stewards. This book is alive in you because it is an autobiography of your fallen state.

May it Be by Enya. Years ago, I read The Lord of the Rings trilogy and I have seen the recent motion pictures. During the writing of this book, I felt compelled to again watch the Lord of the Rings because I now saw many connections to my awakened state. I watched the movies in part at my sister's house. I later felt compelled to purchase my own copy. I purchased a used copy with a dysfunctional episode I had intended to watch. I therefore watched my second choice. As I got up to turn off the movie I heard *May It Be* begin to play. I regard this song as the anthem of this book.

I Ain't Movin by Des'ree. This musical album includes the hit song "You Gotta Be," which dates back to the 1990s. During the writing of this book, I heard this song again on the radio. I felt compelled to purchase the CD. I was overwhelmed by it. I regard this album as the musical soul of this book. "You Gotta Be" and "Crazy Maze" are songs of the quest for Truth. "In My Dreams" is a song of your dream. There are other songs on the album that are inspiring, including, *Herald the Day* and *Love is Here*. I believe the message of this book is more fully experienced with the accompaniment of this album.

"Just Dance" and "Starstruck" by Lady Gaga. I love to sing and dance and I enjoy these dance tracks. The chorus of "Just Dance" says, Just Dance! Gonna be OK, Do do doo doot-n, Just Dance! Spin that record, babe. This is a beautiful expression of your True relationship with Life. If you wish to experience a small taste of the power of Oneness with Life, I suggest playing "Just Dance" and "Starstruck." Don't forget to crank it up!

Jesus Christ Super Star, Universal Pictures, 1973. My older brother took me to see it in the theater when I was twelve. I remember people at my church saying that it should not be viewed because it is heresy. It was somewhat over my head then. However, I have been drawn to this movie and soundtrack ever since I was a teenager. I particularly enjoy the song "Everything's All Right." I have wondered how Jesus, who was portrayed in church as something so perfect and separate from me, could be portrayed as a man with a mission who sometimes did not have all of the answers. I have come to see this movie as highly inspired. It portrays Jesus as an awakened person who lived in a state of Being. However, I believe the movie regards him as more human or limited than he really was. I believe he had great power and could perform miracles.

"Desperado," by The Eagles. Glenn Fry and Don Henley wrote this song in their early days. They were quite young when they wrote it. I see it as inspiring. I have sung this song many times. It advises Desperado to let go of his hardened nature, which will lead him nowhere, and open the gate to something real, such as the queen of hearts, rather than the queen of diamonds. Also, many of The Eagles' later songs are songs of enlightenment.

The Church of Jesus Christ of Latter-day Saints. This church is the entity that is most commonly known as the Mormon Church. This is the church my parents joined in 1952, and the church I was born into in 1960. This church in many ways nurtured me. It gave me the truth, but it also gave me much falseness. The greatest falseness of all is that it projects the ego unto Jesus to make an ego Jesus.

The Book of Mormon. It correctly teaches that truth can be found outside of the Bible. It correctly teaches that the fall of Adam was not contrary to God's will. Other than that it is a tale of the ego, full of contradiction, conflict, and death. If it inspires you, then you are an ego.

Safety for the Soul by Jeffrey Holland. This is a Mormon Conference talk from 2009 and can be seen on YouTube. Holland is a very high official of the Mormon Church. He offers you safety for the soul and he places value on honor. He is an ego offering you the Teddy Bear and he has not seen perfect Love, which knows nothing of honor. He says those who leave the Mormon Church must do so by crawling over, under, or around the *Book of Mormon* to make their exit. I did have to crawl. I crawled for over thirty years, but now I stand tall, and my Father has told me that I never have to crawl again because I am free. Holland's grandfather said of the *Book of Mormon*, "No wicked man could write such a book as this, and no good man would write it unless it were true and he were commanded by God to do so." The people who wrote it were neither wicked nor good. They were simply pious egos that projected themselves onto God and made an idol. This is readily understood if you simply study the workings of the ego.

The Love of God by Dieter Uchtdorf. This is a Mormon Conference talk from 2009 that can be seen on YouTube. Uchtdorf is a very high official of the Mormon Church. His talk is an example of control and subtle deception of a present-day ego religion. He gives what appears to be a very fine sermon on Love and the Love of God, but he states that the Love of God cannot be found with a prayer. He thus attempts to control God as he places God in a little box. Enlightenment or transcendence stories resulting from a single prayer, or no prayer at all, are presented in the Bible, the *Book of Mormon*, and Joseph Smith's first vision story found in the Pearl of Great Price, which is another Mormon scripture. Uchtdorf thus contradicts his own truth, which is characteristic of the ego.

Hinduism. I am not well versed in Hinduism but I have friends that are and they have shared helpful teachings with me. I have read the Bhagavad-Gita, which teaches that nothing is more valued than the presence and assistance of God. Krishna is the voice of transcendence, as is Buddha and Jesus. Krishna and Buddha are of India and lived hundreds of years prior to Jesus. The East is the hemisphere of enlightenment and transcendence and India appears to be the hub of this hemisphere. The Wisemen of the Gospel of Luke are from the East and are of the likeness of Krishna and Buddha. The connection to the East as given in Luke should be reason enough for Christians to look beyond the Bible and explore the teachings of the East. After all, the Wisemen were apparently not Jews or Christians yet if you believe Luke, they are drawn to Judea by the signs of Jesus' birth. How did they manage that? What is the source of their wisdom?

Buddhism. I am a Buddha. Not just a Buddhist. I mean literally a Buddha. I am awakened like Buddha. Buddha means an end to suffering. My pain and suffering has ended. I am full of the Buddha's laughter. I see beyond delusion. I have not transcended to the higher Truths of Siddhartha Gautama, who is the original Buddha but I am on that path. I spent most of my life paying little heed to Buddhism. I saw it as something pagan because this is how I was taught. I now suggest that all people pay close attention to it.

The Wizard of OZ. Some Christians express angst against the Wizard of OZ story because they feel it diminishes the role of Grace. To them, it tells a story of people being self-empowered, rather than God-empowered. For me, it was always a story of empowerment, period, regardless of the source. I did not understand Christian concepts of Grace, but I was still guided and empowered by Grace my whole Life. This indicates that empowerment in itself is a message of salvation. We must keep people moving forward somehow and someday they can fully comprehend the power of Grace. Once you fully comprehend Grace, you realize that there is no boundary between self-empowerment and God-empowerment.

The former Soviet Union (Russia, et al). This is the empire that was the Cold War nemesis of the empire of United States and Western Europe. This resourceful, nation played a critical role in driving the Space race and other technologies that we enjoy today. They allowed the Iron Curtain to fall. This option became increasingly inevitable but it was still a noble move on their part. They could have made the process much more painful. The world owes them our gratitude and respect.

The United States. This nation has blessed the world and me greatly. Some have condemned the United States and have even sought its destruction from within. I say that Lady Liberty stands in New York Harbor for a reason. The United States in many ways is an ideal society. It offers opportunity and equality and it is often generous. It allowed a poor boy like me to be free and receive a quality university education including a master's degree for less than $2000. It is a land of diversity in topography, climate, and people. Its people are Beings of Light. Its people are creative, inventive, and industrious when not overly controlled by the government. The United States has built an empire as the age of empires ended. An empire that stands within the United States today is the federal government. The federal government has become a problem. It has grown into such a monster ego of control that I now long for the days of King George III. The control presently being exercised on the American people goes far beyond a stamp tax or a tea tax. It is a control exercised by the third antichrist. America is the land of the free and the home of the brave. The third antichrist seeks to make America the land of the demon and the home of the slave.

Star Trek or *Star Wars*. These entities are so well known they require no further identification. These are more than entertainment; they are inspired Truth. The grand ideas that exist in human consciousness are placed to serve the purpose of creation. They move us along our path. *Star Trek* or *Star Wars* is our path. It is always our choice to decide if the path is a peaceful one.

How to See Yourself As You Really Are by His Holiness the Dalai Lama. This book discusses many topics, but my favorite aspects are the discussions of the interconnectedness and oneness of all things.

Who Ordered This Truckload of Dung? by Ajahn Brahm. Ajahn is an Australian who is now a senior Buddhist monk. His book is a collection of stories. One of my favorites is about a brick wall that he built. It is an excellent lesson in good enough and perfection. Another favorite part is the three questions: What is the most important time in your Life? Who is the most important person in your Life? What is the most important thing you can do in your Life? When I read the book in the spring of 2009, I was pleased to get the first two answers correct. I now have all three answers correct. The correct answers are now, the person in front of you, and to care. To care means many things. One thing it means is that you must care whether you get all three answers correct.

Why Christianity Must Change or Die by John Shelby Spong, episcopal bishop (Anglican) of Newark, New Jersey. Presents a view of the Bible and historical Christianity from a logical perspective. From this perspective, he identifies many weaknesses. Despite these weaknesses, he still concludes that for some reason that he does not fully comprehend; his heart is drawn to the expression of Jesus Christ. I am pleased to tell John Shelby Spong that his analysis and his heart are True. Christianity must change by recognizing the nature of the ego and teaching the path of enlightenment. After all, that is what Jesus did.

Good to Great by Jim Collins. Promotes greatness through team effort. Everything is a team effort because all is Life. No one is ever separate from Life. Also promotes greatness through an ego-free pursuit of something in which you can be the best in the world. This sounds very much like awakening to your Life's purpose. When you awaken, you are placed on an ego-free pursuit of something that Life itself wishes you to do. By definition, you become the best in the world.

Guide to Investing (What the Rich Invest In, That the Poor and Middle Class Do Not!) by Robert T. Kiyosaki, et al of the Rich Dad's Brand. Teaches that our economic system is rigged against the employee. You are advised to be an entrepreneur or a high-level investor to leverage your time. Truth wishes all to have abundance. Draw close to Truth and all will be added unto you. Your abundance is always personal. When the ego is removed, what you consider to be abundant will likely change dramatically.

Chased Out of Mormonism (The Excommunication of Lyndon Lamborn) by Lyndon Lamborn. Lyndon is a Facebook friend of mine and I have had lunch with him. He was a lifelong dedicated Mormon. He has a technical background and an analytical mind. He started down the path of exploration and discovered the propaganda and deceit within the Mormon Church. He sought Truth and assumed that others within the church were equally interested in Truth. He was mistaken and was excommunicated. This is the fate of many Truth-seeking Mormons. Therefore, those who find that it is not all that it claims to be often conclude that all religion is nothing and turn to atheism. I am happy to say the Mormon Church has some truth but much of what it teaches is false.

Mormons Fighting a Christian, a YouTube video. A street preacher who knows the Biblical Jesus and who has been given a voice by Christ attempts to educate a pair of young Mormon missionaries. The street preacher does not fit in the Mormon box and the missionaries attempt to contend with the unexpected. I say to the missionaries, and the inspired street preacher, that Truth is found everywhere, and often in the most unexpected places.

Currahee! (A Screaming Eagle At Normandy) by Donald R. BurGett. An autobiographical account of a young paratroopers training, and his combat jump into Normandy. He was discouraged when during a training jump he injured his leg and was delayed in the training schedule. The rest of his barracks-mates proceeded with the training schedule and all were killed in a plane crash. If he had not injured his leg, he would have been on the plane that crashed. This story is an example of our inability to judge events as good or bad.

Send Out Cards by Cody Bateman, president of Send Out Cards. Send Out Cards is an Internet-based company that provides a greeting-card service. It provides a website that allows you to select from a catalog of over twelve thousand cards. It also allows you to create your own cards from photos and artwork files. Once you create your card on the computer screen, you simply press the Send button, and the electronic file is sent to a print-house where the card is printed, placed in an envelope, addressed, postmarked, and placed in the mail for you. This gives you a convenient and economical method to send a real card through the postal mail from any computer on the planet. Cody Bateman was inspired to create this company by the death of his older brother Kris. I had often told people that we do not have the wisdom to judge Life, but the death of Kris drove that lesson into me with great power. With that, I was able to love and appreciate Life to a degree that I had never done before. Cody Bateman created Send

Out Cards to promote love and appreciation based on promptings. His belief in promptings has also served to create this book of love and appreciation.

Beach Money by Jordan Adler. A beautiful story of a man who always felt he was destined to become an entrepreneur. His Life changed when he went into the woods with a notebook and wrote the story of his Life as he desired it to be. Some describe this as sending your "I Am statements" to the Genie. His Life became the writing found in his notebook. Jordan, his notebook, and the Genie are the Truth.

In God's Name by CBS. This video can be watched on YouTube. It interviews leaders from twelve major religions. Mostly acknowledges conflict between good and evil, but never offers a solution other than finding a common ground. The hidden stranger within is never mentioned.

Printed in the USA
CPSIA information can be obtained
at www.ICGtesting.com
CBHW020347041224
18174CB00101B/384/J

9 798369 431368